Highway Safety Performance and Statistical Methods 2018

AA005764

Transportation Research Record: Journal of the Transportation Research Board

Volume 2672, Issue 30

ISBN: 978-1-5108-7833-4

Printed from e-media with permission by:

Curran Associates, Inc.
57 Morehouse Lane
Red Hook, NY 12571

proceedings.com

Some format issues inherent in the e-media version may also appear in this print version.

Published by Sage Publications USA

Copyright© (2018) by Transportation Research Board of the National Academies
All rights reserved.

ISBN (Print) 978-1-5108-7833-4
ISBN 2018 Printed Set (All Issues) 978-1-5108-7735-1

Printed by Curran Associates, Inc. (2019)

For permission requests, please contact sagepub.com/journals-permissions

Additional copies of this publication are available from:

Curran Associates, Inc.
57 Morehouse Lane
Red Hook, NY 12571 USA
Phone: 845-758-0400
Fax: 845-758-2633
Email: curran@proceedings.com
Web: www.proceedings.com

Volume 2672 Number 30 December 2018

TRR
JOURNAL OF THE TRANSPORTATION RESEARCH BOARD

Contents

Highway Safety Performance and Statistical Methods 2018

Articles

TRANSPORTATION RESEARCH BOARD 2018 EXECUTIVE COMMITTEE*

OFFICERS

CHAIR: **Katherine F. Turnbull**, *Executive Associate Director and Research Scientist, Texas A&M Transportation Institute, College Station*
VICE CHAIR: **Victoria A. Arroyo**, *Executive Director, Georgetown Climate Center; Assistant Dean, Centers and Institutes; and Professor and Director, Environmental Law Program, Georgetown University Law Center, Washington, DC*
EXECUTIVE DIRECTOR: **Neil J. Pedersen**, *Transportation Research Board*

MEMBERS

Scott E. Bennett, *Director, Arkansas Department of Transportation, Little Rock, AR*
Carlos Braceras, *Executive Director, Utah Department of Transportation, Salt Lake City, UT*
Ginger Evans, *President, Tower Consulting, LLC, Arlington, VA*
Nathaniel P. Ford Sr., *Executive Director–CEO, Jacksonville Transportation Authority, Jacksonville, FL*
A. Stewart Fotheringham, *Professor, School of Geographical Sciences and Urban Planning, Arizona State University, Tempe, AZ*
Susan Hanson, *Distinguished University Professor Emerita, Graduate School of Geography, Clark University, Worcester, MA*
Steve Heminger, *Executive Director, Metropolitan Transportation Commission, San Francisco, CA*
Chris T. Hendrickson, *Hamerschlag University Professor of Engineering, Carnegie Mellon University, Pittsburgh, PA*
Jeffrey D. Holt, *Managing Director, BMO Capital Markets, NY*
S. Jack Hu, *Vice President for Research and J. Reid and Polly Anderson Professor of Manufacturing, University of Michigan, Ann Arbor, MI*
Roger B. Huff, *President, HGLC, LLC, Farmington Hills, MI*
Geraldine Knatz, *Professor, Sol Price School of Public Policy, Viterbi School of Engineering, University of Southern California, Los Angeles, CA*
Melinda McGrath, *Executive Director, Mississippi Department of Transportation, Jackson, MS*
Patrick K. McKenna, *Director, Missouri Department of Transportation, Jefferson City, MO*
Brian Ness, *Director, Idaho Transportation Department, Boise, ID*
James P. Redeker, *Commissioner, Connecticut Department of Transportation, Newington, CT*
Leslie Richards, *Secretary, Pennsylvania Department of Transportation, Harrisburg, PA*
Mark L. Rosenberg, *Executive Director, The Task Force for Global Health, Inc., Decatur, GA*
Gary C. Thomas, *President and Executive Director, Dallas Area Rapid Transit, Dallas, TX*
Pat Thomas, *Senior Vice President of State Government Affairs, United Parcel Service, Washington, DC (Retired)*
James M. Tien, *Distinguished Professor and Dean Emeritus, College of Engineering, University of Miami, Coral Gables, FL*
Dean H. Wise, *Consultant, Dean Wise LLC, Winchester, MA*
Charles A. Zelle, *Commissioner, Minnesota Department of Transportation, Saint Paul, MN*

EX OFFICIO MEMBERS

Ronald Batory, *Administrator, Federal Railroad Administration, Santa Fe, NM*
Mary R. Brooks, *Professor Emerita, Dalhousie University, Halifax, Nova Scotia, Canada, and Chair, TRB Marine Board*
Mark H. Buzby *(Rear Admiral, U.S. Navy), Maritime Administrator, Maritime Administration, U.S. Department of Transportation*
Steven Chalk, *Deputy Assistant for Secretary for Transportation, U.S. Department of Energy*
Steven Cliff, *Senior Advisor to the Chair, California Air Resources Board, Sacramento, CA*
Howard R. Elliott, *Administrator, Pipeline and Hazardous Materials Safety Administration, U.S. Department of Transportation*
Daniel K. Elwell, *Acting Administrator, Federal Aviation Administration, U.S. Department of Transportation*
Audrey Farley, *Associate Administrator for Administration, Pipeline and Hazardous Materials Safety Administration, Washington, DC*
LeRoy Gishi, *Chief, Division of Transportation, Bureau of Indian Affairs, U.S. Department of the Interior, Washington, DC*
John T. Gray II, *Senior Vice President, Policy and Economics, Association of American Railroads, Washington, DC*
Brandye Hendrickson, *Acting Administrator, Federal Highway Administration, U.S. Department of Transportation*
Nikola Ivanov, *Director of Operations, Center for Advanced Transportation Technology Laboratory, University of Maryland, College Park, and Chair, TRB Young Members Council*
Donald Jackson *(Major General, U.S. Army), Deputy Commanding General for Civil and Emergency Operations, U.S. Army Corps of Engineers, Washington, DC*
Heidi King, *Deputy Administrator and Acting Administrator, National Highway Traffic Safety Administration, U.S. Department of Transportation*
Raymond Martinez, *Administrator, Federal Motor Carrier Safety Administration, Washington, DC*
Keith Nelson, *Senior Advisor to the Secretary, Office of the Assistant Secretary for Research and Technology, U.S. Department of Transportation*
Craig A. Rutland, *U.S. Air Force Pavement Engineer, U.S. Air Force Civil Engineer Center, Tyndall Air Force Base, FL*
Karl Schultz *(Admiral, U.S. Coast Guard), Commandant, U.S. Coast Guard*
Karl Simon, *Director, Transportation and Climate Division, U.S. Environmental Protection Agency*
Paul Skoutelas, *President and CEO, American Public Transportation Association, Washington, DC*
Daniel Sperling, *Professor of Civil Engineering and Environmental Science and Policy; Director, Institute of Transportation Studies, University of California, Davis*
K. Jane Williams, *Acting Administrator, Federal Transit Administration, U.S. Department of Transportation*
Frederick G. (Bud) Wright, *Executive Director, American Association of State Highway and Transportation Officials, Washington, DC*

---

* Membership as of August 2018.

TRB

TRANSPORTATION RESEARCH BOARD

CORE SPONSORS

Transportation Departments of the 50 States and the District of Columbia

Federal Government

U.S. Department of Transportation
- Federal Aviation Administration (FAA)
- Federal Highway Administration
- Federal Motor Carrier Safety Administration (FMCSA)
- Federal Railroad Administration (FRA)
- Federal Transit Administration
- National Highway Traffic Safety Administration (NHTSA)
- Office of the Assistant Secretary of Transportation for Research and Technology (OST-R)

U.S. Air Force Civil Engineer Center (AFCEC)
U.S. Coast Guard (USCG)
U.S. Department of Energy
U.S. Department of the Interior (DOI)
U.S. Environmental Protection Agency (EPA)

Nongovernmental Organizations

American Public Transportation Association (APTA)
Association of American Railroads (AAR)
California Air Resources Board

The National Academies of
SCIENCES · ENGINEERING · MEDICINE

The **National Academy of Sciences** was established in 1863 by an Act of Congress, signed by President Lincoln, as a private, nongovernmental institution to advise the nation on issues related to science and technology. Members are elected by their peers for outstanding contributions to research. Dr. Marcia McNutt is president.

The **National Academy of Engineering** was established in 1964 under the charter of the National Academy of Sciences to bring the practices of engineering to advising the nation. Members are elected by their peers for extraordinary contributions to engineering. Dr. C. D. Mote, Jr., is president.

The **National Academy of Medicine** (formerly the Institute of Medicine) was established in 1970 under the charter of the National Academy of Sciences to advise the nation on medical and health issues. Members are elected by their peers for distinguished contributions to medicine and health. Dr. Victor J. Dzau is president.

The three Academies work together as the National Academies of Sciences, Engineering, and Medicine to provide independent, objective analysis and advice to the nation and conduct other activities to solve complex problems and inform public policy decisions. The Academies also encourage education and research, recognize outstanding contributions to knowledge, and increase public understanding in matters of science, engineering, and medicine.

Learn more about the National Academies of Sciences, Engineering, and Medicine at **www.national-academies.org.**

The **Transportation Research Board** is one of seven major programs of the National Academies of Sciences, Engineering, and Medicine. The mission of the Transportation Research Board is to increase the benefits that transportation contributes to society by providing leadership in transportation innovation and progress through research and information exchange, conducted within a setting that is objective, interdisciplinary, and multimodal. The Board's varied activities annually engage about 7,000 engineers, scientists, and other transportation researchers and practitioners from the public and private sectors and academia, all of whom contribute their expertise in the public interest. The program is supported by state transportation departments, federal agencies including the component administrations of the U.S. Department of Transportation, and other organizations and individuals interested in the development of transportation.

Learn more about the Transportation Research Board at **www.TRB.org**.

Article

TRR
JOURNAL OF THE TRANSPORTATION RESEARCH BOARD

Transportation Research Record
2018, Vol. 2672(30) 1–8
© National Academy of Sciences:
Transportation Research Board 2018
Article reuse guidelines:
sagepub.com/journals-permissions
DOI: 10.1177/0361198118758054
journals.sagepub.com/home/trr
SAGE

Safety Evaluation of the SafetyEdge Treatment for Pavement Edge Drop-Offs on Two-Lane Rural Roads

Craig Lyon[1], Bhagwant Persaud[2], and Eric Donnell[3]

Abstract

This study estimated crash modification factors (CMFs) for the SafetyEdge paving technique that is applied for the treatment of pavement edge drop-offs on two-lane rural highways. An empirical Bayes observational before-after evaluation based on installation data in Iowa, North Carolina, Ohio, Pennsylvania, and Florida found that the SafetyEdge treatment was associated with statistically significant reductions in fatal and injury (FI), run-off-the-road (ROR), opposite-direction (head-on and sideswipe), and drop-off-related crashes. The ROR CMF was also statistically significant for both horizontal curved and tangent sections. A disaggregate analysis examined the variation of the CMF for ROR with factors such as the travel lane width, traffic volume, presence of a horizontal curve, posted speed limit, and the pre-treatment ROR crash frequency. The results of that analysis indicated, for example, that the SafetyEdge paving technique appears to have a greater ROR safety benefit on two-lane rural roadway segments with average annual daily traffic volumes greater than 3,000 vehicles per day, relative to roadway segments with lower traffic volumes. A crash modification function (CMFunction) was calibrated with expected pre-treatment ROR crashes as the independent variable to simultaneously capture the relationship of the CMF for ROR crashes to multiple factors. An economic analysis found that the treatment is highly cost-effective.

Over half of all fatalities in the United States involve a roadway departure, defined by FHWA as a crash that occurs after a vehicle crosses an edge line, a center line, or otherwise leaves the traveled way. Pavement edge drop-offs are known to contribute to some of the most severe roadway departure crashes. The typical scenario begins with one or more of the vehicle tires leaving the paved roadway. If the driver attempts to return to the paved surface without slowing, and a significant vertical edge drop-off is present, the sidewall of one or more tires will scrub against this edge and the driver may oversteer in an effort to return to the paved roadway. This may result in a driver losing control of the vehicle, contributing to a head-on collision with a vehicle traveling in the opposing direction, a rollover, or a run-off-the-road (ROR) event onto the roadside adjacent to the opposing traveled way. To mitigate roadway departure crashes, it is important to reduce the number of vehicles that encroach onto the roadside, and to minimize the consequences of a roadway departure event. One strategy to achieve the second objective is to eliminate vertical pavement edges that may become drop-offs over the life of the pavement.

The SafetyEdge is constructed with a low-cost paver attachment that enables the pavement edge to be paved and compacted to a finished 30° angle to promote a safe return to the travel lane after a roadway departure (*1*). Figure 1 shows a typical pavement cross-section with the SafetyEdge for a pavement-resurfacing project. After paving, the SafetyEdge is backfilled and graded flush with the paved surface ("New Graded Material" in Figure 1). Where the compacted material covering the pavement SafetyEdge settles or erodes, the angled edge is easily traversable by vehicles attempting to re-enter the roadway when compared with a vertical or near-vertical pavement edge drop-off.

Review of Literature

A SafetyEdge evaluation by Graham et al. considered treatment sites on rural two-lane highways with paved and unpaved shoulders in Georgia and Indiana (*2*). The treatment sites were those re-paved with the SafetyEdge treatment, while the comparison sites were defined as those similar to the treatment sites that were resurfaced without the SafetyEdge. An empirical Bayes (EB) observational before-after approach was used to assess the safety effects of the treatment. A total of 77 treatment sites, with a combined length of nearly 182 miles, were included in the EB observational before-after study. Seventy-five comparison sites, totaling 212 miles, were also included in the sample. The

[1]Persaud & Lyon, Inc., Ottawa, ON, Canada
[2]Department of Civil Engineering, Ryerson University, Toronto, ON, Canada
[3]Department of Civil and Environmental Engineering, Pennsylvania State University, University Park, PA

Corresponding Author:
Address correspondence to Craig Lyon: craig.lyon @rogers.com

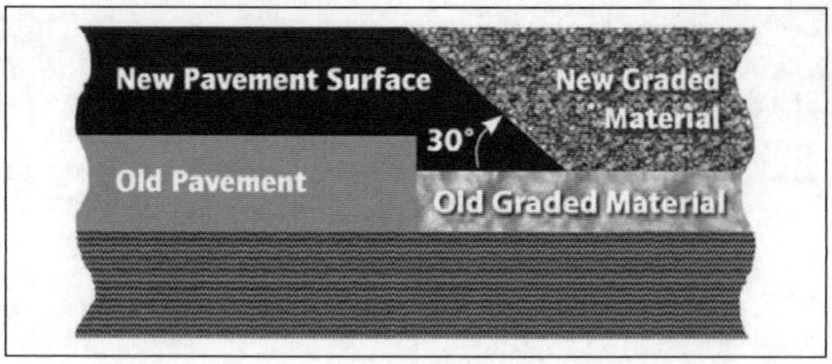

Figure 1. Illustration of SafetyEdge (*1*).

before period consisted of 4 years of data, while the after period consisted of 3 years of data. A crash modification factor (CMF) for the SafetyEdge was derived by computing the ratio between a CMF for the SafetyEdge treatment and a CMF for the comparison group. The results showed that the SafetyEdge treatment is associated with a 5.7% reduction in total crashes, a 6.3% reduction in ROR crashes, and a 5.6% reduction in drop-off-related crashes on two-lane rural highways; however, the results were not statistically significant. Cross-sectional regression models using only treatment and comparison site data after re-paving two-lane highways produced similar results to the EB evaluation. The authors reported —cost–benefit ratios ranging from 4 to 63 resulting from installation of the SafetyEdge treatment on two-lane rural highways, depending on the shoulder type (paved or unpaved) and traffic volume.

A recent study in Iowa (*3*) considered the safety effectiveness of the SafetyEdge treatment on rural two-lane highways with posted speed limits of 45 miles per hour (mph) or higher. Sites that were re-paved with the SafetyEdge in 2010 and 2011 were compared with the safety performance of control segments, which were generally adjacent and similar to the SafetyEdge locations. The EB methodology was used to estimate CMFs for total and target crashes, which were defined as ROR crashes occurring to the left or right of the roadway. The CMFs for total crashes were 0.87, 0.84, and 0.90 for all severities, injury crashes only, and unknown and property damage only (PDO) crashes, respectively. All of the CMFs for total crashes were statistically significant at the 90% confidence level. The CMFs for target crashes were 0.88, 0.92, and 0.86 for all severities, injury crashes only, and unknown and PDO crashes, respectively. Only the injury CMF for target crashes was not statistically significant for the target crash type.

There are some issues that have limited the applicability of the results of these previous SafetyEdge evaluations and motivated the current study. These include the following:

- Graham et al. did not examine the potential effects on the CMF of various roadway and roadside features (e.g., lane and shoulder width, shoulder type, horizontal curve presence and direction, and roadside hazard rating) (*2*). Inclusion of these variables in a disaggregate safety evaluation may help identify site-specific features that benefit more (or less) from the SafetyEdge treatment. The sample size was small, which not only restricted the ability to conduct a disaggregate analysis, but also produced an overall EB safety-effect estimate that was not statistically significant (*2*).
- The Hallmark et al. study could not be considered definitive for national applicability in that it pertained to one state and, even so, only data for roadways with speed limits of 45 mph or more were included (*3*). As well, the study did not examine the potential effects on the CMF of various roadway and roadside features.

This study builds on these efforts by using a multi-state database to estimate the effect of SafetyEdge on crashes on rural two-lane roadways using state-of-the-art methodology. A large sample of re-paving projects with and without the SafetyEdge afforded the opportunity to develop CMFs for several crash types, and to disaggregate the CMFs based on traffic volume and other roadway features. The sample also enabled consideration of a crash modification function (CMFunction) to capture variability in the safety effect, and the undertaking of an economic analysis to evaluate the cost-effectiveness of this treatment.

Data Collection

Data from the states of Iowa, North Carolina, Ohio, Pennsylvania and Florida, were included in the evaluation. Sites that were re-paved with or without the SafetyEdge in 2012 or earlier were included in the treatment and reference group samples respectively. The analysis period ranged from 2005 through 2015, depending on the treatment installation dates and availability of data from the participating state transportation agencies. At least 3 years of before and after period data were sought for each treatment and reference group site. Only re-paving projects that were completed during a single calendar year were included in the analysis, so this year was excluded from the safety evaluation. Table 1 summarizes the data elements and sources of data, while Table 2 summarizes the number treatment and reference group sites, mile-years of data, average daily traffic volumes (AADT), and total crashes for the before and after periods. Mile-years of data is the sum of the length of a segment multiplied by the number of years of crash data summed over all sites.

Intersection-related and animal crashes were excluded from the data. Total, fatal+injury (FI), ROR, and opposite-direction crashes (OPPDIR) could be identified in all of the crash data files supplied by the five agencies included in the evaluation. Opposite-direction crashes included both head-on (HO) and sideswipe-opposite-direction (SSOD) crashes. Drop-off-related crashes could be identified in the Iowa,

Table 1. Data Elements and Information Sources

Data Description	Elements	Information Source(s)
Crash Data	Location Type (non-intersection, intersection etc.) Crash type Severity Crash sequence Date of crash Crash location (e.g., route/milepost)	Electronic crash reporting system
Exposure and Speed Limit Data	Average annual daily traffic Segment length Posted speed limit	Electronic roadway inventory files
Roadway Cross-section Data	Lane width Shoulder type and width	Electronic roadway inventory files
Horizontal Alignment Data	Curve presence Direction of curve Radius and length of curve (if available)	Electronic roadway inventory files Video photolog review Web-based review
Roadside Hazard Rating	Subjective roadside hazard rating (on a scale of 1 to 7)	Video photolog review
Pavement Edge Drop-off Data	Depth and shape of edge drop-off data	Field measurement
Treatment Installation Dates	Dates of SafetyEdge installation at treatment sites Dates of resurfacing only at reference sites	Outreach to STAs
Cost of SafetyEdge	Unit costs per mile of applying SafetyEdge	Outreach to STAs

Note: STA = state transportation agency.

Table 2. Safety Evaluation Data Summary

Agency	Site Type	Number of Sites	Miles	Time Period	Mile-Years	Mean AADT	Total Crashes
Iowa	Treatment	279	316.1	Before	1,008.1	933	393
				After	1,204.9	1,756	308
	Reference	108	74.2	Before	211.1	1,321	96
				After	283.8	2,002	102
Marion County, Florida	Treatment	40	36.4	Before	164.9	6,250	130
				After	126.6	5,675	162
	Reference	51	21.3	Before	99.3	3,616	101
				After	71.4	3,546	104
North Carolina	Treatment	727	369.1	Before	2,374.6	3,285	2,496
				After	947.6	3,065	1,139
	Reference	573	196.8	Before	1,262.5	2,059	847
				After	501.6	2,275	397
Ohio	Treatment	447	576.0	Before	3,846.6	3,535	3,899
				After	605.7	2,998	718
	Reference	163	203.0	Before	1,295.9	3,185	1,168
				After	286.3	2,442	272
Pennsylvania	Treatment	170	24.0	Before	72.1	1,493	53
				After	72.1	1,560	38
	Reference	1,411	159.0	Before	318.4	1,382	277
				After	636.7	1,370	503

Note: AADT = annual average daily traffic volumes.

North Carolina, Ohio, and Pennsylvania data files. This crash type was defined as one in which either (a) the roadway contributing circumstance variable was coded to indicate that the shoulder was low, soft, or high, or (b) the sequence of events indicated a ROR right event followed by either a ROR left, overturn/rollover, or cross centerline/median event. In Pennsylvania, the sample of drop-off-related crashes was too small to develop a CMF for this crash type. In Iowa, drop-off-related crashes represented 2.5% to 13.5% of the total crashes during the before or after periods at the treatment or reference group sites. In North Carolina and Ohio, drop-off-related crashes ranged from 7.2% to 13.9% and 3.5% to 6.6%, respectively, of the total crashes.

Methodology

The EB before-after approach is currently accepted as state-of-the-art in observational before-after studies of crash data (*4*). The EB analysis properly accounts for regression-to-the-mean, if it exists, and for differences in traffic volume and for crash trends between the periods before and after treatment. The approach is comprised three basic steps, each defined as follows:

- Step 1: Predict what the safety performance of rural two-lane highways would have been in the after period had the SafetyEdge not been implemented.
- Step 2: Estimate what the actual safety performance was in the after period with the SafetyEdge.
- Step 3: Compare the results of Step 1 and Step 2.

In approaching Step 1, the reference group was used to account for the effects of traffic volume changes and temporal effects on safety due to the variation in weather, demographics, and crash reporting. This was done through the calibration and application of safety performance functions (SPFs) relating crashes of different types and severities to traffic flow and other relevant factors for each jurisdiction. This enabled the simultaneous accounting for temporal and possible regression-to-the-mean effects, as well as those related to changes in traffic volume. The SPF used the reference group sites, which included two-lane rural highways that were resurfaced without the SafetyEdge. In calibrating the SPFs, annual factors were included that not only reflected time trends but also the effects of resurfacing without SafetyEdge over time. This approach separates out the effects of resurfacing from the effects of the SafetyEdge.

Data required for SPF development included crash, traffic volume, and geometric data as well as information on resurfacing and the application or non-application of SafetyEdge for the sample of sites that were resurfaced. These data were required for each year of the analysis period. Generalized linear modeling (GLM) was used to fit the SPF parameters with generalized estimating equations (GEE). The GEE procedure is used when unknown correlations, such as temporal correlations in annual crash data, is present, as is the case here since each site-year was used as a separate observation in the analysis file. GLM allows the specification of various error structures, including the negative binomial error structure that is generally recognized as being appropriate for modeling crash data.

The expected number of crashes had no treatment been applied, B, used the SPF to first estimate the number of crashes that would be expected in each year of the before period at locations with traffic volumes and other characteristics similar to a treatment site being analyzed. The sum of these annual SPF estimates (P) was then combined with the count of crashes (x) in the before period at the treatment site to obtain an estimate of the expected number of crashes (m) before the treatment. This estimate of m is

$$m = w(P) + (1 - w)(x) \tag{1}$$

The weight (w) was estimated as follows:

$$w = 1/(1 + kP) \tag{2}$$

where: k is the estimated overdispersion parameter of the assumed negative binomial distribution for the crash counts used in estimating the SPF.

A factor was then applied to m to account for the length of the after period and differences in traffic volumes between the before and after periods. This factor is the sum of the annual SPF predictions for the after period divided by P, the sum of these predictions for the before period. The result, after applying this factor, is an estimate of B. The procedure also produced an estimate of the variance of the expected number of crashes that would have occurred in the after period without the treatment.

The estimate of B is then summed over all sites in a treatment group of interest (to obtain B_{sum}) and compared with the count of crashes during the after period in that group (A_{sum}). The variance of B is also summed over all sections in the group of interest.

The index of safety effectiveness (θ) was estimated as

$$\theta = (A_{sum} / B_{sum}) / \{1 + [Var(B_{sum}) / B_{sum}^2]\} \tag{3}$$

The standard deviation (SD) of θ is

$$Stddev(\theta) = [\theta^2\{[Var(A_{sum}) / A_{sum}^2] + [Var(B_{sum}) / B_{sum}^2]\} / [1 + Var(B_{ssm}) / B_{sum}^2]^2]^{0.5} \tag{4}$$

The percent change in crashes is $100(1-\theta)$; thus, a value of $\theta = 0.70$ with an SD of 0.12 indicates a 30% reduction in crashes with an SD of 12%.

A disaggregate analysis was also conducted to examine how site conditions may affect the value of the CMF. This was accomplished by isolating sites that had certain levels of a given variable and estimating the separate effects for each level. In the case of continuous variables, such as traffic volume, aggregation occurred over specified ranges of the variable. The estimation of a CMFunction to capture the variation of the CMF was also undertaken.

The following independent variables were considered, either in developing the SPFs for the EB analysis, or for disaggregating the results of the evaluation:

- Traffic volume (average annual daily traffic, veh/day)
- Segment length (miles)
- Lane width (feet)
- Shoulder width (feet)
- Pavement width (feet)
- Horizontal curve presence (yes/no)
- Horizontal curve radius and length (if available)
- Roadside hazard rating (1 = clear roadside; 7 = roadside hazards present near roadway)

Analysis and Results

Approximately 1,321 miles of rural two-lane highways re-paved with the SafetyEdge were included in the safety evaluation. The reference group included nearly 654 miles of two-lane rural highways re-paved without the SafetyEdge. The reference group data from each state transportation agency was used to estimate separate SPFs for total, fatal + injury (FI), ROR, head-on+sideswipe-opposite-direction (HO+SSOD) and drop-off related 0crashes. The SPFs were then used in the EB analysis to develop CMFs for each crash type for each state.

The SPFs that were developed using data from Florida are shown below for illustration and were of the form

$$\text{Total Crashes per year} = \text{length} * \exp^{(\text{Intercept}+\text{yrpstresurf}+b*\text{alignmnent})}\text{AADT}^{c} \quad (5)$$

$$\text{FI Crashes per year} = 0.624 * \text{length} * \exp^{(\text{Intercept}+\text{yrpstresurf}+b*\text{allignment})}\text{AADT}^{c} \quad (6)$$

$$\text{ROR Crashes per year} = 0.346 * \text{length} * \exp^{(\text{Intercept}+\text{yrpstresurf}+b*\text{aalignmen})}\text{AADT}^{c} \quad (7)$$

$$\text{OPPDIR Crashes per year} = 0.039 * \text{length} * \exp^{(\text{Intercept}+\text{yrpstresurf}+b*\text{alignment})}\text{AADT}^{c} \quad (8)$$

where: length = segment length in miles; yrpstresurf = a variable to represent the number of years post-resurfacing, which is a constant term to account for time trends and effects of years after resurfacing the alignment = 1 if segment is a tangent and 0 if a horizontal curve; and AADT = average annual daily traffic in vehicles per day. Note that drop-off related crashes could not be identified in Florida and as such no SPF was developed. The statistical modeling output for the Florida SPFs is shown in Table 3.

The yrpstresurf variable accounts for both time trends and the effects of resurfacing on safety, both immediately after treatment and over time. When applying the SPFs, the variable yrpstresurf includes five levels, which represents the number of years post-resurfacing. A value of 0 indicates a year prior to resurfacing and the values 1 through 4 represent the years post-resurfacing. For the year of interest, the SPF is applied using

Table 3. Florida Safety Performance Functions

Regression Parameter	Estimate	Standard Error
Intercept	−10.3466	3.8712
Yrpstresurf	0: −0.4513	0.3453
	1: 0.1530	0.4140
	2: −10.7500	0.4939
	3 0.1544	0.4265
	4: 0.0000	0.0000
b (alignment)	−0.7908	0.2758
c (AADT)	1.3597	0.4622
Overdispersion	0.9300	0.2707

Note: AADT = annual average daily traffic volumes.

the value of the estimate in Table 3. For example, for the first year post-resurfacing, the value 0.1530 is entered into the SPFs.

Aggregate Results

Table 4 shows the aggregate results of the EB before-after evaluation for all states combined for the crash types of interest. The CMF for all states combined was derived after confirming that the CMF estimates were reasonably consistent amongst states.

For total crashes, the CMF is very close to 1.0, i.e. no effect, and is statistically insignificant. This was not unexpected as the SafetyEdge treatment is only expected to affect crashes involving a vehicle running off the road and not the majority of crashes.

The CMF for FI crashes, by contrast, is statistically significant and less than 1.0, indicating that the SafetyEdge is associated with fewer FI crashes. This was also not unexpected, because the events involved in a pavement edge drop off crash tend to be severe.

For ROR crashes, the CMF is statistically significant and less than 1.0 indicating that the SafetyEdge reduces these crashes by improving the chances that a run-off-road vehicle can safely re-enter the roadway.

The CMF for OPPDIR crashes is statistically significant indicating that the SafetyEdge paving technique is associated with fewer crashes of these types, some of which result from drivers overcorrecting after one or more tires drop off the edge of the pavement.

For the drop-off-related crashes the CMF is statistically significant and less than 1.0, although data from Florida or Pennsylvania could not be used because drop-off-related crashes could not be explicitly identified, or the sample of such crashes was small.

Disaggregate Results

The objective of the disaggregate analysis was to assess whether meaningful conclusions could be drawn from an investigation of the variability of the CMFs under differing application scenarios. Such insights could be meaningful in themselves, but they could also inform the development of a CMFunction to capture this variability mathematically.

Table 4. Aggregate Results

Data Element	Crash Type				
	Total	FI	ROR	OPPDIR	Drop-Off Related
EB estimate of crashes expected in the after period without SafetyEdge	2,391.14	1,069.08	564.50	190.09	175.52
Count of crashes in the after period	2,365	954	446	155	115
Estimate of CMF	0.989	0.892*	0.790*	0.813*	0.655*
Standard error of estimate of CMF	0.025	0.034	0.042	0.076	0.064

Note: FI = fatal and injury; ROR = run-off-the-road; OPPDIR = opposite direction crashes; EB = empirical Bayes; CMF = crash modification factor.
*Statistically significant at 95% confidence level.

For the disaggregate analysis, a univariate categorical analysis was conducted to estimate CMFs for various categories or ranges of variables logically of interest for which data were consistently available across sites and states, and for which preliminary analysis indicated that meaningful insights could be obtained. These variables included speed limit, surface width of travel lanes, average AADT in the before period, and expected crash frequency per mile before treatment, all of which are likely substantially correlated. In addition, it was desired to examine the effect on the CMF of presence of a curve in a segment for the states that provided relevant information.

The final analysis was performed using data for all states combined by summing the terms in equations 3 and 4 for all sites, and was focused on ROR crashes, one of the target crash types, after initial analysis for total crashes did not reveal any meaningful insights. This was likely because there were no overall effects on total crashes, with the effects on ROR crashes diluted since they constitute only about 20% of all crashes. Drop-off related and FI crashes were considered, but the sample was much too small to provide meaningful conclusions. Similarly, for OPPDIR crashes, the sample was too small even to provide additional insights if included with ROR, especially since an unknown portion of those crashes may not be drop-off related.

Table 5 presents the results of the univariate categorical analysis for the ROR CMFs based on site characteristics. These results indicate that the CMF is smaller at a higher crash rate, which was expected. The CMF is statistically significant and indicative of a safety benefit for all total traveled way widths (the sum of the two-lane widths) categories greater than 18 ft. There are indications that the benefit is more pronounced on roads with wider travel-lane widths. The ROR CMF is statistically significant for both tangent and curved roadway segments with the benefit being more pronounced for tangent segments relative to segments with horizontal curves. This is likely because it is easier to recover from a roadway departure along a tangent segment of roadway than along a horizontal curve. The ROR CMFs for the two categories of posted speed limit are both statistically significant and less than 1.0 with the numbers suggesting that the SafetyEdge is associated with fewer ROR crashes on two-lane rural highways with posted speed limits equal to or less than 55 mph. Finally, the results indicate that the benefit of SafetyEdge for ROR crashes is greater on roads with higher AADTs.

It should be emphasized that these are univariate categorical analysis results and that CMF differences between any two categories may also be reflective of differences in values of other variables. For example, the differences between travel lane width categories may also reflect differences in effects related to AADT as is evident from the results in Table 5. Given the evident correlative effects, a CMFunction was developed using a single independent variable, the expected crash frequency per mile-year before treatment, which logically captures the effects of the other variables investigated in the univariate categorical analysis. In developing the CMFunction, the use of state as a variable in the model or as a random effect variable did not improve the CMFunction. The CMFunction, which was estimated using linear regression, is

$$CMF_{ROR} = 0.975 - \left(\begin{array}{c} 0.432 \times \text{Expected ROR crash} \\ \text{frequency per mile-year} \end{array} \right) \quad (9)$$

To apply this CMFunction, the expected ROR crash frequency per mile-year can be estimated using an SPF for ROR crashes for two-lane rural roads that are calibrated to the jurisdiction in question. If the site has a reported crash history, the estimate can be improved by applying the EB calculation, which combines the SPF estimate with the observed count of ROR crashes per year. Caution should be exercised in applying the CMFunction beyond the limit of estimation data (2.85 expected ROR crashes/mile-year).

Economic Analysis

An economic analysis was conducted whereby the safety benefits were derived based on the estimated CMF of 0.790 for ROR crashes (See Table 4). For the benefit calculations, the most recent FHWA mean comprehensive crash costs from Council et al., which were developed using 2001 cost information, were used as a base (*5*). These were updated to 2016 dollars considering changes to the U.S. Department of

Table 5. Disaggregate Analysis for ROR Crashes for Various Site Features

Feature	Category	Sites	CMF	Standard Error	Average AADT Before	Crashes Observed After
EB expected crashes/mile/ year before treatment	<0.34[1]	1,379	1.009	0.075	2,560	220
	>0.34	284	0.651*	0.049	4,464	219
Traveled way width with different break points	16-18 ft.	253	0.981	0.118	1,366	80
	19-24 ft.	1,166	0.777*	0.046	3,082	357
	16-20 ft.	683	0.866*	0.066	1,782	201
	21-24 ft.	737	0.764*	0.057	3,695	236
	16-22 ft.	908	0.881*	0.055	2,250	330
	23-24 ft.	512	0.643*	0.069	3,707	107
Alignment	All tangent	519	0.759*	0.069	2,989	154
	Part or all curve	417	0.836*	0.060	2,176	230
Speed limit	55 mph	1,049	0.818*	0.048	2,812	341
	<55 mph	335	0.695*	0.094	3,958	74
AADT	<3000[1]	1,027	0.917	0.069	1,286	223
	>3000	636	0.692*	0.052	5,468	223

Note: AADT = annual average daily traffic volumes.
[1]Change point selected to achieve a similar number of after-period crashes in each category.
*Statistically significant at 95% confidence level.

Transportation (USDOT) value of a statistical life (VSL) from 2001 to 2016 (*6*). The resulting aggregate 2016 unit cost for a ROR crash was obtained as US$313,667. Considering the annual crash reduction based on the CMFs, an annual benefit of US$16,612,854 for ROR crashes was estimated.

The material considered in the construction cost was hot-mix asphalt (HMA) with an average cost of US$75.00 per ton for an HMAsurface or wearing course from the five transportation agencies included in the evaluation. For an assumed 10-year service life, a discount rate of 7% based on the Office of Management and Budget Circular A-4 (*7*), and information on typical SafetyEdge cross-sections, the annualized SafetyEdge treatment cost was estimated to range from US$10.65 per mile-year to US$21.30 per mile-year for 2-in. and 4-in. HMA depths, respectively. For the 1,321.63 miles of two-lane rural highways re-paved with the SafetyEdge in the five states in the present evaluation, the total annual costs are US$28,151 and US$14,075, respectively, for the 4-in. and 2-in. deep-paved roadways.

The benefit–cost ratio (B/C), calculated as the ratio of the annual benefit to the annual cost, is 590 and 1,180 for the 4-in. and 2-in. depth assumptions, respectively. The USDOT recommends that sensitivity analysis be conducted by assuming values of a statistical life 0.56 and 1.40 times the recommended 2016 value (*6*). These factors can be applied directly to the estimated B/C ratio to get a range of 330 to 826 for the 4-in. depth assumption and 661 to 1652 for the 2-in. depth assumption. This suggests that, even with conservative assumptions on service life and the VSL, the SafetyEdge can be applied cost-effectively.

Conclusions

This study estimated CMFs for the SafetyEdge treatment on two-lane rural highways using data from five transportation agencies. An EB observational before-after evaluation found statistically significant CMFs aggregated across all states for FI, ROR and opposite-direction crashes. A statistically significant CMF was also found for drop-off related crashes, although the data were only available from three states. The total crash CMF, aggregated across all states, was 0.989. This result was not statistically significant at the 95% confidence level, which was somewhat expected because the SafetyEdge is intended to mitigate a subset of roadway departure crashes. The fatal+injury CMF was 0.892 for all states combined, and was statistically significant at the 95% confidence level. This result was expected, because the SafetyEdge addresses crash types that often result in severe outcomes. The ROR and drop-off-related CMFs were 0.790 and 0.655, respectively, and were statistically significant for all states combined. The CMF for OPPDIR crashes of 0.813 was also statistically significant. All of these crash types are related to roadway departure events, which the SafetyEdge treatment is intended to mitigate.

Disaggregate CMFs were also estimated for two-lane rural highways with the SafetyEdge. This univariate analysis found that the ROR CMF is lower when the ROR crash frequency is higher before the SafetyEdge is applied on a repaving project. The ROR CMF is statistically significant and less than 1.0 for paved roadway surfaces that are at least 19-ft wide. The SafetyEdge treatment appears to be associated with a greater ROR safety benefit on two-lane rural roadway segments that have posted speed limits less than 55 mph than on those with 55 mph speed limits; the benefit is also greater for tangent sections than for curved sections and for segments with average annual daily traffic volumes greater than 3,000 vehicles per day, relative to roadway segments with lower traffic volumes. A CMFunction was developed for ROR crashes based on the expected ROR crash frequency prior to treatment, which logically captures the correlative effects of the other variables investigated in the univariate categorical analysis.

An economic evaluation found that the SafetyEdge paving technique can be applied cost-effectively based on the CMF for ROR crashes. Even with conservative estimates of the pavement service life, VSL, and the depth of the pavement surface (i.e., construction costs), the B/C ratio was large, ranging from 330 to 1,652.

In summary, this research indicates that the SafetyEdge is a highly cost-effective safety countermeasure based on the aggregate CMFs developed in the present study for the ROR crash type, and based on the expected reduction in FI crashes at the study locations.

Acknowledgments

This work was managed by Dr. Carol Tan of the Federal Highway Administration. Additional technical guidance was offered by Cathy Satterfield and Christopher Wagner from FHWA. The authors wish to acknowledge their vision and guidance to the project. This work is based on data provided by Iowa, North Carolina, Ohio, Pennsylvania, and Florida. The support of the state transportation agencies in providing data and advice is gratefully acknowledged.

References

1. Federal Highway Administration (FHWA). *The SafetyEdge: A Pavement Edge Drop-off Treatment*. Publication Number FHWA-SA-10-034, 2010.
2. Graham, J. L., K. R. Richard, M. K. O'Laughlin, and D. W. Harwood. *Safety Evaluation of the SafetyEdge Treatment. Report No. FHWA-HRT-11-024*, Federal Highway Administration, McLean, VA, March 2011.
3. Hallmark, S. L., A. Goswamy, and M. D. Pawlovich. *Safety Edge Crash Modification Factors*. Iowa State University Digital Repository, Tech Transfer Summaries. Iowa State University, Institute for Transportation, Ames, IA, August 2016.
4. Hauer, E. *Observational Before-After Studies in Road Safety – Estimating the Effect of Highway and Traffic Engineering Measures on Road Safety*. Elsevier Science, Inc., Tarrytown, New York, 1997.
5. Council, F., E. Zolashnja, T. Miller, and B. Persaud. *Crash Cost Estimates by Maximum Police-Reported Injury Severity Within Selected Crash Geometrics, FHWA-HRT05-051*, 2005.
6. U.S. Department of Transportation. *Guidance on Treatment of the Value of a Statistical Life (VSL) in U.S. Department of Transportation Analyses—2016 Adjustment*. Memo, August 8, 2016. https://www.transportation.gov/sites/dot.gov/files/docs/VSL%20Guidance%202016.pdf. Accessed Dec. 19, 2016.
7. Office of Management and Budget. *Circular A-4*. Found at: https://www.whitehouse.gov/omb/circulars_a004_a-4/. Accessed Aug. 24, 2016.

The Standing Committee on Highway Safety Performance (ANB25) peer-reviewed this paper (18-00673).

TRR
JOURNAL OF THE TRANSPORTATION RESEARCH BOARD

Article

Transportation Research Record
2018, Vol. 2672(30) 9–20
© National Academy of Sciences:
Transportation Research Board 2018
Article reuse guidelines:
sagepub.com/journals-permissions
DOI: 10.1177/0361198118767409
journals.sagepub.com/home/trr
SAGE

Development of Safety Performance Functions: Incorporating Unobserved Heterogeneity and Functional Form Analysis

Behram Wali[1], Asad J. Khattak[1], Jim Waters[2], Deo Chimba[3], and Xiaobing Li[1]

Abstract

To improve transportation safety, this study applies Highway Safety Manual (HSM) procedures to roadways while accounting for unobserved heterogeneity and exploring alternative functional forms for Safety Performance Functions (SPFs). Specifically, several functional forms are considered in Poisson and Poisson-gamma modeling frameworks. Using 5 years (2011–2015) of crash, traffic, and road inventory data for two-way, two-lane roads in Tennessee, fixed- and random-parameter count data models are calibrated. The models account for important methodological concerns of unobserved heterogeneity and omitted variable bias. With a validation dataset, the calibrated and uncalibrated HSM SPFs and eight new Tennessee-specific SPFs are compared for prediction accuracy. The results show that the statewide calibration factor is 2.48, suggesting rural two-lane, two-way road segment crashes are at least 1.48 times greater than what HSM SPF predicts. Significant variation in four different regions in Tennessee is observed with calibration factors ranging between 2.02 and 2.77. Among all the SPFs considered, fully specified Tennessee-specific random parameter Poisson SPF outperformed all competing SPFs in predicting out-of-sample crashes on these road segments. The best-fit random parameter SPF specification for crash frequency includes the following variables: annual average daily traffic, segment length, shoulder width, lane width, speed limit, and the presence of passing lanes. Significant heterogeneity is observed in the effects of traffic exposure-related variables on crash frequency. The study shows how heterogeneity-based models can be specified and used by practitioners for obtaining accurate crash predictions.

As core tools provided in Highway Safety Manual (HSM), published by the American Association of State Highway Officials (AASHTO), Safety Performance Functions (SPFs) are used to estimate expected crash frequencies at a particular facility. Importantly, SPFs that can accurately predict crashes are valuable to state departments of transportation (DOTs) as they help in identifying areas with potential safety concerns. Crash predictive models are used to statistically estimate the expected crash frequency for particular facility type with specified "base" conditions (*1*). The default values of key facility characteristics are used in developing HSM SPFs. If base conditions are not met, HSM provides crash modification factors (CMFs) that can be multiplied with base-case predictions from HSM SPF to capture the differences between base and jurisdiction-specific conditions. Since crash frequency and associated under- and over-dispersion in crash data can vary significantly across jurisdictions, AASHTO clearly recognizes the need for calibrating HSM SPFs to specific jurisdictions (*1*). Compared with uncalibrated counterparts, calibrated HSM predictive models can provide more accurate crash predictions by accounting for variations in crash data and associated factors. However, when enough data are available, HSM allows and encourages transportation agencies to develop jurisdiction- or state-specific SPFs in obtaining more accurate crash predictions.

Calibration of HSM predictive models and development of new SPFs have been performed by several states throughout the US (*2–4*). By using localized data, jurisdiction-specific SPFs are developed by regressing crash frequency as a function of certain explanatory factors. While jurisdiction-specific SPFs (compared with HSM SPFs) can better represent local

[1]Department of Civil & Environmental Engineering, The University of Tennessee, Knoxville, TN
[2]Strategic Transportation Investments Division, Tennessee Department of Transportation, Nashville, TN
[3]Civil & Architectural Engineering, Tennessee State University, Nashville, TN

Corresponding Author:
Address correspondence to Asad J. Khattak: akhattak@utk.edu.

conditions, traffic crash frequencies and associated factors (e.g. traffic volumes) can still vary significantly across similar, or even identical, road geometry and conditions within a jurisdiction (*5*). The correlations between crashes and associated factors can be heterogeneous, and it is important to correct for heterogeneity in modeled relationships that can arise from several observed and unobserved factors relating to (but not limited to) (*5–9*):

- Driver behaviors
- Vehicle types
- Socioeconomic factors
- Traffic and pavement characteristics
- Road geometrics
- Variations in police accident recording thresholds
- Other time and space related unobserved factors.

For a complete review of methodological challenges in crash frequency modeling, interested readers are referred to Lord and Mannering (*10*).

As part of a research project performed by researchers at the University of Tennessee to facilitate the implementation of new HSM procedures in the state of Tennessee, the main objectives of this study are to apply HSM predictive models for rural two-lane, two-way roads, compute calibration factors, and explore the need for developing Tennessee-specific SPFs. Before conducting a detailed empirical analysis of rural two-lane, two-way road safety, crash rates are analyzed as an effective "*first brush*" tool to quantify relative safety in different regions within Tennessee. Methodologically, several functional forms based on HSM guidelines and other forms prevalent in literature are considered. Importantly, the development of rigorous SPFs undertaken in this study accounts for unobserved heterogeneity. The models can be used for more accurate prediction of crashes. Finally, out-of-sample prediction forecasts are generated to evaluate the performance of all competing models. Consideration of unobserved heterogeneity in the development of SPFs is critical, as ignoring heterogeneity in the effects of explanatory factors can lead to inconsistent, biased estimates. Although this important methodological concern is adequately recognized and extensively addressed in extant safety literature, such methods are rarely used in practice by state DOTs for more accurate prediction of crashes. Notably, the scope of this study is limited to accounting for random heterogeneity as opposed to the possibility of systematic heterogeneity due to non-linearity.

Methodology

This study involves the following tasks: crash rate analysis, calibrating HSM SPF models to better represent local conditions, and development of jurisdiction-specific SPFs. Different techniques employed to achieve the tasks are briefly explained in this section.

Crash Rates

Crash rate is commonly used for safety evaluation of roadway facilities in practice. In this analysis, crash rates are estimated using two different measures of exposure, that is, crash rates by vehicles miles traveled (VMT), and crash rates by segment length (*1*). Specifically, crash rate by VMT is calculated using Equation 1 (*1*):

$$R = \frac{C*100,000,000}{V*365*N*L} \tag{1}$$

where: R = Crash rate per VMT;
C = Total crashes in study period (5 years);
V = Traffic volume using Annual Average Daily Volumes (AADT);
N = Number of years of data; and
L = Length of roadway segment in miles.

Likewise, crash rates by segment length is calculated using Equation 2 (*1*):

$$R = \frac{C}{N*L} \tag{2}$$

where: R = Crash rate per mile of segment;
N = Number of years of data; and
L = Length of roadway segment in miles.

Calibration Factor Estimation

SPF for rural two-lane, two-way road segments in HSM is developed using data from selected states in the US. Equation 3 represents rural two-lane, two-way SPF, and can be applied when certain base conditions (as documented in HSM) meet local jurisdiction conditions to which HSM SPF is applied (*1*):

$$N_{\text{SPF}} = AADT*L*365*10^{-6}*e^{-0.312} \tag{3}$$

where: N_{SPF} = Base predicted number of crashes in study period for site i;
L = Length of roadway segment (miles); and
$AADT$ = Annual average daily traffic during study period.

SPF in Equation 3 is estimated for specific base conditions related to roadway geometric features. Whenever base-case conditions are not met, CMFs can be used that account for the differences in local jurisdiction-specific geometric features and base conditions assumed in HSM SPF. That is, crashes obtained using Equation 3 can be multiplied with CMFs as

$$N = N_{\text{SPF}} \times CMF_1 \times CMF_2 \times CMF_3 \times \ldots\ldots\ldots \times CMF_i \tag{4}$$

where: N = Adjusted predicted crash frequency in local jurisdiction;
CMF_i = Crash modification factor(s) for road segment features deviating from HSM base conditions.

After calculating predicted crash frequency (accounting for non-base jurisdiction-specific conditions), the calibration factor can simply be calculated using Equation 5 (*1*):

$$C = \frac{\sum Observed\ Crashes}{\sum Adjusted\ Predicted\ Crashes} = \frac{N_{\text{observed}}}{N} \quad (5)$$

where: N_{observed} = Observed crash frequency in study period; and N = Adjusted predicted crash frequency in local jurisdiction.

Finally, the calculated calibration factor (C) in Equation 5 can be multiplied with HSM SPF (Equation 3) for predicting rural two-lane, two-way road segment crashes in a specific jurisdiction. The resulting SPF then becomes

$$N = C * \left(AADT * L * 365 * 10^{-6} * e^{-0.312} \right) \quad (6)$$

In what follows, uncalibrated HSM SPF (Equation 3) will be referred to as Model 1, whereas HSM SPF with calibration (Equation 6) referred to as Model 2.

Tennessee-Specific Safety Performance Functions

When enough data are available, it is recommended that users develop jurisdiction-specific SPFs (*1*). In addition to better crash forecasts, developing state-specific SPFs can help in network screening and evaluation of engineering treatments at a site or project level. Given the discrete non-negative data nature of crashes, count data modeling techniques are typically used to model crash frequencies as a function of explanatory variables. Common techniques include Poisson generalized linear models (GLMs) and negative binomial GLMs (*10*).

Poisson and Negative Binomial Regressions. For Poisson models, the probability of having a specific number of crashes "*n*" at road segment "*i*" is written in Equation 7 (*10*):

$$P(n_i) = \frac{\exp(-\lambda_i)\lambda_i^n}{n_i!} \quad (7)$$

where: $P(n_i)$ = probability of crash occurring at segment "*i*," "*n*" times per specific time-period; and λ_i = Poisson parameter for segment "*i*" which is numerically equivalent to segment "*i*" expected crash frequency per year $E(n_i)$.

Formally, λ_i can be viewed as a log link function of a set of explanatory factors using Equation 8 (*10*):

$$\ln(\lambda_i) = \beta(X_i) \quad (8)$$

where: X_i = Vector of explanatory variables; and β = Vector of estimable parameter estimates.

Poisson function defined in Equations 7 and 8 can be maximized by the standard maximum likelihood procedure (*11*). Application of Poisson regression to over-dispersed crash data can result in inappropriate results. If the mean and variance of crash data are not equal, corrective measures are applied to Equation 8 by adding an independently distributed error term, ε, as:

$$\ln(\lambda_i) = \beta(X_i) + \varepsilon_i \quad (9)$$

where: $\exp(\varepsilon_i)$ = Gamma-distributed error term with mean one and variance α (*12*).

Following Khattak et al. (*13*), if α is statistically significantly different from zero, negative binomial regression should be favored, otherwise the Poisson model is more appropriate.

Unobserved Heterogeneity in Crash Frequency Modeling. As discussed earlier, it is likely that correlations between key explanatory variables and crash frequency may not be consistent across multiple rural two-lane, two-way segments. There are several compelling reasons to expect intrinsic unobserved heterogeneity (*11,14,15*). For instance, data used for crash frequency analysis usually have a limited set of variables and other variables influencing crash frequencies are usually not available. Furthermore, if key variables are omitted from analysis and too few variables are included in modeling, parameter estimates may be biased and inaccurate. One way to address this issue is to allow parameter estimates to vary across observations (*15*). As such, random parameters can be included in estimation framework as (*14,15*):

$$\beta_i = \beta + \varphi_i \quad (10)$$

where: φ_i = randomly distributed term with any pre-specified distribution (e.g. normal distribution) with mean zero and variance σ^2 (*15*).

With Equation 10, Poisson parameter in Equation 8 becomes:

$$\lambda_i \mid \varphi_i = EXP(\beta X) \quad (11)$$

In addition, the Poisson parameter in Equation 9 in Poisson-Gamma model becomes:

$$\lambda_i \mid \varphi_i = EXP(\beta X + \varepsilon_i) \quad (12)$$

Finally, likelihood function for a random-parameter model can be maximized through the maximum simulated likelihood using Equation 13 (*11*):

$$LL = \sum_i ln \int_{\varphi_i}^{i} g(\varphi_i) P(n_i|\varphi_i) d\varphi_i \quad (13)$$

where: $g(.)$ = Probability density function of randomly distributed term with pre-specified distribution (e.g. normal distribution) with mean zero and variance σ^2.

Different distributions are tested for random parameters such as normal, log-normal, uniform, triangular, and tent. However, density function based on normal distribution resulted in best fit. For simulation purposes, 200 Halton draws are used. More details on random parameter models can be found in the literature (*10,15*).

Functional Forms. To develop Tennessee-specific SPFs, different functional forms based on original HSM form and functional forms used by other researchers are considered using both Poisson and Negative Binomial regression techniques (*3,13*).

First Form (Models 3 and 4). Functional forms for Tennessee-specific SPFs in Models 3 and 4 are similar to HSM base SPF (Equation 3) in terms of simple structure and minimal data requirements. Thus, model 3 (Poisson distribution) and model 4 (negative binomial distribution) are of the form:

$$N_{\text{TN-SPF}} = AADT * L * 365 * 10^{-6} * e^{\beta_o} \tag{14}$$

Poisson and negative binomial regressions based on functional form in Equation 14 are equivalent to constant-only regression models with exposure as an offset variable. Exposure is calculated as $AADT * L * 365 * 10^{-6}$ (*1*).

Second Form (Models 5 and 6). Second functional form tested is similar to HSM functional form (Equation 13) in terms of minimal data requirements. For example, only AADT and segment length are used as two potential explanatory variables in Models 5 (Poisson regression) and 6 (negative binomial regression). However, functional form differs from HSM functional form (Equation 13) in terms of variable specification, and is:

$$N_{\text{TN-SPF}} = exp(\beta_o) * AADT^{\beta_1} * Seglen^{\beta_2} \tag{15}$$

Where β_o, β_1, and β_2 are parameters to be estimated. Functional form in Equation 15 is equivalent to regressing crash frequency on natural logarithm of AADT and segment length, and is used by several researchers in modeling crash frequencies (*4,11,13*).

Third Form (Models 7 and 8). Third functional form is using Equation 16 (*10*):

$$\ln(N_{\text{TN-SPF}}) = \beta_o + \sum_{i=1}^{P} \beta_i X_i \tag{16}$$

Where: X_i= Matrix of explanatory factors; and β_i = Column-vector of parameter estimates associated with each variable in matrix X_i.

Model 7 refers to SPF based on Poisson regression whereas model 8 is based on negative binomial regression.

Fourth Form (Models 9 and 10). Finally, Model 9 incorporates unobserved heterogeneity in crash data in addition to all available variables as in Models 7 and 8. This can potentially provide insights into heterogeneous effects of different factors on crash frequency. SPFs in Models 9 and 10 are developed using simulation based random parameter Poisson and negative binomial modeling techniques. The functional form is

$$N_{\text{TN-SPF}} = \exp^{[\beta_o + \sum_{i=1}^{Q} \gamma_i Z_i + \sum_{i=1}^{P} \beta_i X_i]} \tag{17}$$

where: Z_i= Matrix of explanatory factors of random parameters;

γ_i = Column-vector of parameter estimates associated with each variable in matrix Z_i;

X_i= Matrix of explanatory factors of fixed parameters; and

β_i = Column-vector of parameter estimates associated with each variable in matrix X_i.

Model Validation and Goodness-of-fit Measures

One major objective of estimating more realistic crash frequency models is to enhance crash forecast accuracy. To compare estimated models in terms of model fit and out-of-sample predictions, data were randomly divided into two categories, one for model training and the other for model testing/validation. Specifically, 70% of data is used for model fitting. All models (Models 3–10) are estimated and fitted with training data. The remaining 30% of data (N = 90) is used for model testing/validation of HSM SPFs (Models 1–2) and Tennessee-specific SPFs (Models 3–10). To quantify uncertainties in out-of-sample predictions, mean absolute error (MAE) (*12*), root mean square error (RMSE), and mean prediction bias (MPB) are calculated (*12*). MAE and RMSE give the average magnitude of variability in predictions, that is, smaller values are preferred. Unlike MAE and RMSE, MPB can be positive or negative. A positive value of MPB indicates SPF is overestimating number of crashes, whereas a negative value implies underestimation. For evaluating goodness-of-fit and statistical adequacy of all models: log-likelihood at convergence, McFadden R^2, Akaike information criteria (AIC), and Bayesian information criteria (BIC) are used (*12,16*). A lower value of AIC and BIC indicates a relatively better model.

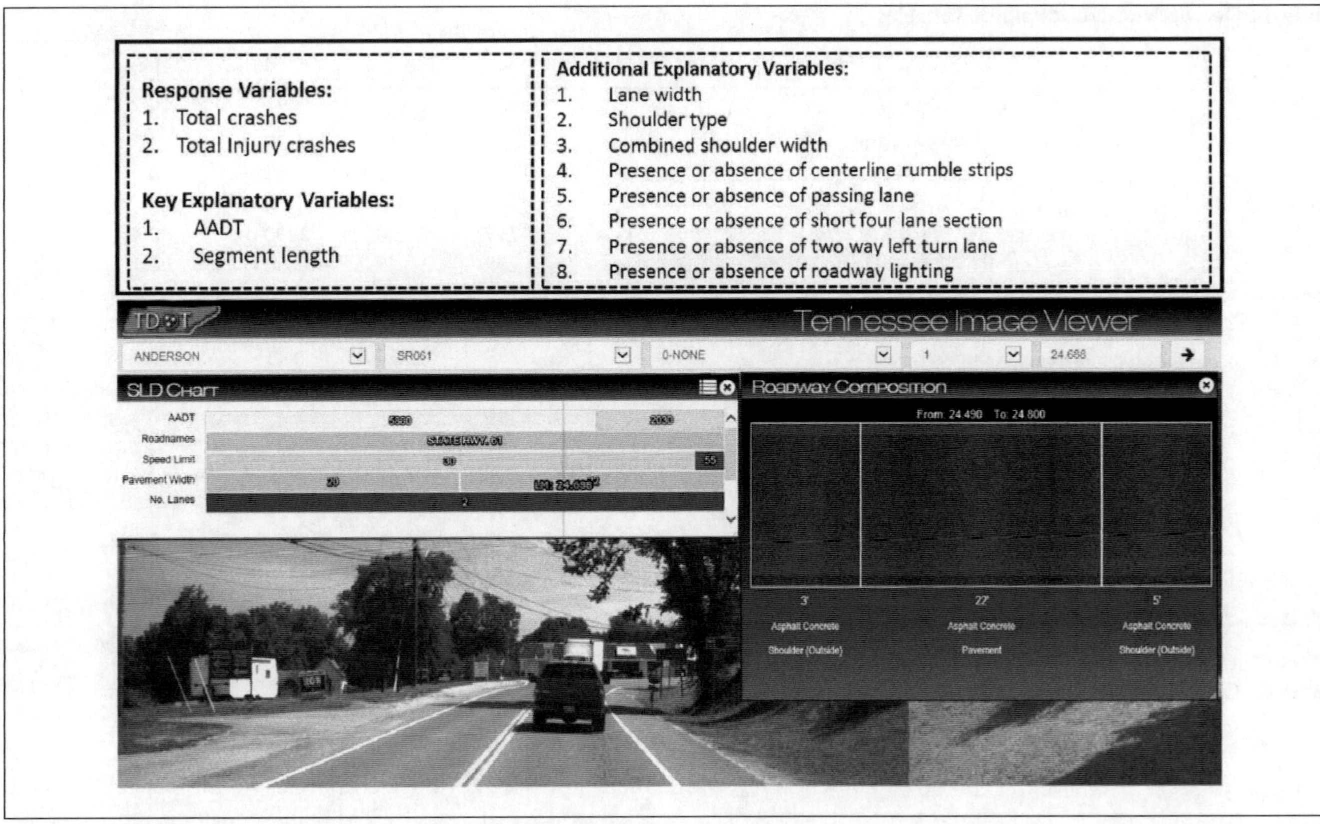

Figure 1. Illustration of E-TRIMS and Tennessee image viewer software for manual extraction of key roadway geometric data.

Data Assembly

To meet study objectives, significant efforts went into assembling data from different sources manually. Specifically, a statewide crash and roadway inventory database is used to calibrate HSM SPFs and develop Tennessee-specific new rural two-lane, two-way road segments SPFs. Crash, traffic, and roadway geometrics data are collected for a 5-year period (2011–2015). Five years of crash and geometric data are used for both calibration of HSM SPFs and development of Tennessee-specific SPFs. For calibration factor analysis, it is conducted both for each year and an average of 5 years of data. For data collection, inventory data on rural two-way, two-lane roadway segments across Tennessee (with a minimum segment length of 0.10 miles) were collected and compiled from the Tennessee Department of Transportation's (TDOT) Enhanced Tennessee Roadway Information Management System (E-TRIMS) (*17*). A total of 14,777 road segments were identified. Then, a random sample of 299 homogeneous roadway segments with complete data was obtained from original 14,777 sample. The random sample accounts for diverse geographical conditions across Tennessee with 61 segments from Region 1 (Knoxville area), 61 segments from Region 2 (Chattanooga area), 104 segments from Region 3 (Nashville area), and 72 segments from Region 4 (Memphis area). Next, 5-year crash data for the selected roadway segments were extracted manually from E-TRIMS through crash summary reports. To ensure the accuracy of manually extracted data, a computer program was used to assign crashes to each of the 299 roadway segments. Doing so revealed successful and accurate matching of crashes for the sampled 299 roadway segments. As shown in Figure 1, roadway inventory data (e.g. segment length) and traffic data (e.g. AADT and speed limits) were also manually extracted from TDOT's digital image viewer (*17*) and traffic count data program (*18*).

To account for omitted variable bias, data on additional correlates of crashes, as recommended by HSM (*1*), were also collected. Note that data on some of these variables could not be used because of unavailability, including horizontal and vertical alignment, roadside hazard rating, and automated speed enforcement. Notably, the heterogeneity model specifications account for omitted variable biases that may arise as a result of missing information on important correlates (*5,11*).

Results

Descriptive Statistics

Table 1 presents descriptive statistics of crash frequencies on rural two-lane, two-way roads in Tennessee, key and additional variables. Based on their distributions, key summary

Table 1. Descriptive Statistics of Key Variables

	Variable	*n*	Mean	Std. dev.	Min	Max
Crash frequencies	Total 5 years crashes	299	6.234	9.413	0	73
	Average 5 years crashes	299	1.247	1.883	0	14.6
	Average 5 years crashes (rounded to nearest integer)	299	1.230	1.951	0	15
Key variables	Total 5 years AADT	299	9141.0	11453.0	302	73056
	Average of 5 years AADT	299	1828.0	2290.0	60.4	14611.2
	Segment length (miles)	299	1.149	1.303	0.1	7.2
Additional variables	Lane width (feet)	299	10.436	1.258	7	12
	Combined shoulder width (feet)	299	3.348	2.593	0	12
	Gravel shoulder	299	0.291	0.455	0	1
	Asphalt concrete shoulder	299	0.421	0.495	0	1
	Turf shoulder	299	0.288	0.453	0	1
	Lighting present	299	0.241	0.428	0	1
	Speed limit (miles per hour)	299	39.866	9.562	20	55
	Presence of C/L rumble strips	299	0.187	0.399	0	1
	Presence of passing lane	299	0.268	0.443	0	1
	Presence of short four-lane section	299	0.010	0.100	0	1
	Presence of two-way left-turn lane	299	0.017	0.128	0	1

Table 2. Crash Rates by VMT and Segment Length

Crash rates/100-million VMT					
Area	*n*	Mean	Std. dev.	Min	Max*
All Regions	299	287.62	384.19	0	3320.88
Region 1	61	331.93	390.97	0	2265.16
Region 2	61	309.09	533.22	0	3320.88
Region 3	104	282.93	331.28	0	1570.97
Region 4	73	239.34	291.81	0	1331.77
Crashes per each mile of roadway per year					
Area	*n*	Mean	Std. dev.	Min	Max
All Regions	299	1.759	3.793	0	33.99
Region 1	61	2.100	4.817	0	31.43
Region 2	61	1.351	1.646	0	7.00
Region 3	104	2.247	4.885	0	33.99
Region 4	73	1.121	1.681	0	9.82

Note: (*) the extremely high maximum crash rates (as indicated by "max" column) are for segments with low number of crashes but also very short segments, i.e. ~ 0.15 *miles*; *n* = sample size; Std. dev. = standard deviation.

statistics, and extraction from a well-organized and integrated state database, underlying data are of reasonable quality. For example, average 5 years of crashes across 299 rural two-lane, two-way segments are distributed with a mean of 1.247 and standard deviation of 1.883, which highlights slight over-dispersion in crash data. Regarding key variables, the mean of average 5 years of AADT is 1828, whereas mean segment length is observed to be 1.149 miles. Lighting is present on approximately 24% of these roads, whereas center line rumble strips and passing lanes are present on approximately 19% and 27% of roadways, respectively. Descriptive statistics of other variables can be interpreted similiarly.

Crash Rates by Vehicles Miles Traveled and Segment Length

Table 2 provides average crash rates for all four regions and regionwide crash rates by VMT and segment length. Average crash rate per 100-million VMT for all regions is 287.62 with a standard deviation of 384.19. Based on simple crash rates, Region 4 appears to be the safest with a crash rate of 239.34 crashes/100-million VMT, followed by Region 3 with 282.93 crashes/100-million VMT.

Calibration Factors Results

Two types of calibration factors are calculated, (1) Base Calibration Factors (CF_{Base}), and (2) Adjusted Calibration Factors (CF_{Adj}). CF_{Base} are estimated by applying HSM SPF (Equation 3) on AADT and segment length only. CF_{Adj} are estimated by applying Equation 4 in estimating crashes while incorporating CMFs for cases where Tennessee-specific roadway geometrics deviate from HSM default values. The calibration procedure is applied for all regions combined as well as separately for each region. In addition, to account for temporal variations, the calibration procedure is both applied for an average of 5-year data, as well as separately for each year. However, the year-wide calibration factors did not exhibit significant variations.

CF_{Base} and CF_{Adj} for the entire state is found to be 2.980 and 2.489, respectively, showing that rural two-lane, two-way road segment crashes are at least 1.48 times greater than what HSM SPF predicts even after accounting for non-base Tennessee-specific conditions. The calibration factor ranged between 2.023 and 2.776 for four regions in Tennessee. For Region 1 and 3, CF_{Adj} is estimated to be 2.584 and 2.776, respectively. For Region 2 and 4, they are 2.444 and 2.023

Table 3. Goodness-of-Fit Statistics of TN-Based SPFs

Functional Form	TN-Specific SPFs	n	LL(Null)	LL (Convergence)	DF	AIC	BIC
First form*	Model 3	209	−233.022	−233.022	1	468.04	471.39
	Model 4	209	−232.39	−232.39	2	468.79	475.48
Second form**	Model 5	209	−410.68	−227.4	3	460.80	470.83
	Model 6	209	−325.37	−227.39	4	462.78	476.15
Third form***	Model 7	209	−410.68	−256.03	7	526.06	549.46
	Model 8	209	−325.37	−261.03	8	527.31	551.60
Fourth form****	Model 9	209	−466.91	−246.6	9	509.31	541.10
	Model 10	209	−466.9135	−246.608	10	513.22	546.64

Note: n = sample size; LL(Null) = log-likelihood of constant-only model; LL(Convergence) = log-likelihood at convergence; DF = number of parameters estimated; AIC = Akaike information criteria; and BIC = Bayesian information criteria.

*First form is $N_{TN\text{-}SPF} = AADT * L * 365 * 10^{-6} * e^{\beta_o}$;

**Second form is $N_{TN\text{-}SPF} = exp(\beta_o) * AADT^{\beta_1} * Seglen^{\beta_2}$;

***Third form is $\ln(N_{TN\text{-}SPF}) = \beta_o + \sum_{i=1}^{P} \beta_i X_i$;

****Fourth form is $N_{TN\text{-}SPF} = \exp[\beta_o + \sum_{i=1}^{Q} \gamma_i Z_i + \sum_{i=1}^{P} \beta_i X_i]$

respectively. Overall, calibration factor results suggest that observed crashes on Tennessee rural two-lane, two-way roads are significantly greater than crashes predicted by calibrated HSM SPFs.

Tennessee Specific Safety Performance Functions

In this section, results of Tennessee-specific SPFs are presented. Specifically, eight different SPFs are developed based on different distributional assumptions (Poisson and negative binomial) and different functional forms. All fixed parameter models (Models 3–8) are estimated via standard maximum likelihood procedures, whereas random-parameter models (Models 9–10) are estimated via simulated maximum likelihood procedures. All models are derived from a systematic process to include most important variables on basis of statistical significance, specification parsimony, and intuition. First, a series of ordinary least square regressions were estimated to spot correlations and patterns in the data. Next, a series of Poisson and negative binomial regressions (both fixed and random parameter) were estimated. Specifically, all variables from Table 1 were tested. Finally, statistically significant variables retain in final model specifications.

Model Selection and Performance Comparison. Before detailed discussion of Tennessee-specific SPFs, goodness-of-fit measures of all estimated models with different distributional assumptions and functional forms are presented in Table 3. Following Washington et al. and Bozdogan (*12,16*), BIC and AIC can be used to evaluate competing nested and non-nested models. Among all estimated models, Model 5 (Poisson SPF based on logarithms of AADT and segment length) has the lowest AIC and BIC indicating relatively superior "in-sample" fit, followed by Model 3 (Poisson SPF based on HSM functional form), and finally, Model 9 (Random Parameter Poisson SPF). Note that goodness-of-fit measures (e.g. log-likelihood at convergence and likelihood-ratio test statistic) presented in Table 3 are indicators for "explanatory" power of competing models. Importantly, "explaining" versus "predicting" are two different dimensions for which statistical models are estimated (*19*). While AIC is derived from a predictive viewpoint, it is an indicator of "in-sample" fitting capabilities of competing models (*19*), and not an indicator of "out-of-sample" forecasts accuracy. In other words, a model may have high out-of-sample forecast errors with lower AIC.

Although SPFs based on only two covariates (e.g. logarithms of AADT and segment length in Models 5–6) perform relatively best, it is interesting to note that Tennessee-specific SPF based on HSM functional form (which only includes estimation of intercept) performs relatively well in fitting "training" data. SPFs based on including all covariates (third form) have the highest AIC and BIC values (lower in-sample fit).

Modeling Results. Tables 4 and 5 present results of all Tennessee-specific SPFs for rural two-lane, two-way roads. Specifically, Table 4 summarizes results of fixed parameter SPFs based on Poisson and negative binomial distributions, and different functional forms, whereas Table 5 summarizes results of random parameter Poisson and Negative Binomial SPFs. Referring to parameter estimates in Tables 4 and 5 for different models, a positive sign on parameter estimate shows that specific variable is positively correlated with crash frequency, and vice versa. For instance, in Model 5, AADT and segment length are positively correlated with crash frequency. This finding is in agreement with the extant traffic safety

Table 4. Estimation Results (Tennessee-Specific Fixed Parameter Count Data Models) (Models 3–8)

	Model 3[a]		Model 4[b]		Model 5[a]		Model 6[b]		Model 7[a]		Model 8[b]	
	β	*t*-stat	β	*t*-stat	β	*t*-stat	β	*t*-stat	β	*t*-stat	β	*t*-stat
Constant	0.7468	12.36	0.774	10.93	−5.456	−10.89	−5.462	−10.78	−1.765	−4.99	−1.949	−5.05
Exposure*(offset)	1	—	1	—	—	—	—	—	—	—	—	—
AADT (ln form)	—	—	—	—	0.783	12.17	0.784	12.03	—	—	—	—
Segment length (ln form)	—	—	—	—	0.904	14.35	0.904	14.26	—	—	—	—
AADT (in thousands)	—	—	—	—	—	—	—	—	0.271	10.41	0.291	7.822
Segment length	—	—	—	—	—	—	—	—	0.456	12.41	0.511	10.23
Shoulder width	—	—	—	—	—	—	—	—	−0.165	−4.73	−0.172	−4.21
Speed limit	—	—	—	—	—	—	—	—	0.024	2.81	0.023	1.21
LW dummy (1 if lane width ≥ 10)	—	—	—	—	—	—	—	—	0.075	0.37	0.169	0.73
Passing lane dummy	—	—	—	—	—	—	—	—	0.203	1.29	0.225	1.17
Over-dispersion	—	—	0.0515	0.856	—	—	0.005	0.119	—	—	0.191	2.3
Summary Statistics**												
McFadden-R^2	0		0		0.446		0.301		0.375		0.025	
χ^2 Statistic[c]	—		—		366.56		195.96		309.29		151.89	
Prob > Critical χ^2	—		—		0.0000		0.0000		0.0000		0.0000	
N	209		209		209		209		209		209	

Note: (*) exposure is AADT*segment length*365*0.000001; (**) other in-sample goodness-of-fit statistics are provided in Table 3; β = parameter estimate; (*a*) = SPF based on Poisson distribution; (*b*) = SPF based on negative binomial distribution; (*c*) = no Chi-square statistic is applicable for models 3 and 4 as they are constant-only models where all other coefficients are forced to be zero; (—) = not-applicable.

Table 5. Estimation Results (Tennessee-Specific Random Parameter Count Data Models) (Models 9 and 10)

	Model 9[a]			Model 10[b]		
Variables	β	SE	*t*-stat	β	SE	t-stat
Fixed parameters						
Constant	−1.9	0.314	−6.04	−1.9	0.321	−5.91
Shoulder width	−0.166	0.033	−5.02	−0.166	0.033	−4.96
Speed limit	0.016	0.007	2.01	0.018	0.009	1.97
LW dummy (1 if lane width >= 10, 0 otherwise)	0.335	0.106	3.16	0.321	0.202	1.58
Passing lane dummy	0.266	0.131	2.031	0.264	0.149	1.789
Random parameters						
AADT (in thousands)	0.289	0.027	10.52	0.278	0.025	11.12
standard deviation	0.036	0.015	2.37	0.035	0.011	3.181
Segment length	0.543	0.043	12.56	0.541	0.041	13.19512
standard deviation	0.16	0.025	6.2	0.17	0.026	6.53846
Over-dispersion	—	—	—	0.003	0.07	0.052
Summary statistics*						
McFadden-R^2		0.471			0.469	
χ^2 statistic		440.60			438.12	
Prob > critical χ^2		0.0000			0.0000	
N		209			209	

Note: (*) other in-sample goodness-of-fit statistics are provided in Table 3; β = parameter estimate; SE = standard error; (*a*) = random-parameter SPF based on Poisson distribution; (*b*) = random-parameter SPF based on negative binomial distribution; (—) = not-applicable.

literature (*3,4,13*). Likewise, in Model 7, shoulder width is negatively associated with crash frequency. Overall, results shown in Table 4 suggest the absence of over-dispersion, which favors statistical superiority of Poisson regression-based SPFs. This finding can also be confirmed in Table 3, in which AICs and BICs of Poisson-based SPFs are almost equal to AICs and BICs of their negative binomial counterparts.

Coming to results in Table 5, random parameter Poisson and negative binomial SPFs resulted in better statistical fit

Table 6. Out-of-Sample Forecast Evaluation

Functional Forms	Model	Mean absolute error	Root mean square error	Mean prediction bias
HSM SPFs	HSM-SPF (Model 1)	2.009	1.417	−0.710
	(HSM-SPF)* CF_{Adj} (Model 2)	1.615	1.270	−0.049
First Form	$N_{TN\text{-}SPF} = AADT * L * 365 * 10^{-6} * e^{\beta_o}$			
	TN-SPF- Poisson Regression (Model 3)	**1.569**	**1.252**	**0.18**
	TN-SPF- NB Regression (Model 4)	1.606	1.267	0.179
Second Form	$N_{TN\text{-}SPF} = exp(\beta_o) * AADT^{\beta_1} * Seglen^{\beta_2}$			
	TN-SPF- Poisson Regression (Model 5)	**1.265**	**1.125**	**0.051**
	TN-SPF- NB Regression (Model 6)	1.266	1.125	0.05
Third Form	$\ln(N_{TN\text{-}SPF}) = \beta_o + \sum_{i=1}^{P} \beta_i X_i$			
	TN-SPF- Poisson SPF (Model 7)	2.208	1.485	0.354
	TN-SPF- NB SPF (Model 8)	3.327	1.824	0.515
Fourth Form	$N_{TN\text{-}SPF} = \exp[\beta_o + \sum_{i=1}^{Q} \gamma_i Z_i + \sum_{i=1}^{P} \beta_i X_i$			
	TN-Specific Random Parameter Poisson SPF (Model 9)	**1.199**	**1.095**	**0.034**
	TN-Specific Random Parameter Negative Binomial SPF (Model 10)	1.224	1.315	0.041

Note: Highlighted in bold are models that perform better in terms of out-of-sample predictions.

relative to fixed parameter counterparts as in Models 7–8 in Table 4, as is shown by smaller AIC and BIC values in Table 3 as well. Specifically, random parameter estimation technique helps in capturing heterogeneous associations between response outcome and explanatory variables. For instance, in random parameter Poisson regression (Model 9), both AADT and segment length are observed to be normally distributed random parameters, suggesting that these variables' effects vary across different rural two-lane, two-way road segments in Tennessee. It is observed that increase in shoulder width is associated with smaller crash frequency. Contrarily, increase in speed limit, lane width, and presence of passing lanes are associated with increase in crash frequency. Incorporating presence of significant unobserved heterogeneity results in significantly better AIC and BIC values, and McFadden R^2 (0.471 for random parameter Poisson compared with 0.375 for fixed parameter Poisson). Also, results in Table 5 suggest that random parameter Poisson regression is statistically superior to random parameter negative binomial regression. This is intuitive, as negative binomial regression models are typically used to capture over-dispersion in crash data. With random-parameter Poisson regressions, it is likely that majority of over-dispersion is captured in the form of unobserved heterogeneity and thus no (or little) over-dispersion may be left in data for a random parameter negative binomial-based regression model to capture (*20*).

Out-of-Sample Forecast Evaluation. While goodness-of-fit statistics presented in Table 5 provide valuable insights regarding "in-sample" fit of all estimated models, "out-of-sample" forecast accuracy cannot be readily inferred from statistics presented in Table 3. As such, a model's true "out-of-sample" forecast capabilities are evaluated by using holdout set technique and calculating statistics that help evaluate "out-of-sample" forecast capabilities of all estimated models. Results of out-of-sample forecast errors of all estimated models are presented in Table 6. Smaller values of MAE and RMSE are desirable. Positive MPB values show that SPF is overestimating, and negative values show that particular SPF is underestimating. Among all models tested, Model 9 (TN-Specific Random Parameter Poisson SPF) exhibited the best out-of-sample forecast capabilities with lowest MAE, RMSE, and MPB, followed by Model 5, and Model 3.

Finally, "out-of-sample" mean-estimated over mean-observed number of crashes for HSM SPFs (with and without calibration), Tennessee-specific Poisson SPF with logarithms of AADT and segment length as explanatory variables (Model 5), and random parameter Poisson SPF with all covariates included (Model 9) are shown in Figure 2. Compared with uncalibrated HSM SPF, "out-of-sample" forecasts for calibrated HSM are closer to the mean equivalence line (red line in Figure 2). Also, compared with HSM SPFs (calibrated and uncalibrated) and Tennessee-specific Poisson SPF based on logarithms of AADT and segment length only, the out-of-sample predictions for Tennessee-specific random parameter Poisson SPF are more evenly distributed across the mean equivalence line (Figure 2). Given all statistical evidence, random parameter-based Poisson SPF can be used in practice in Tennessee and can facilitate generation of more accurate crash predictions. The

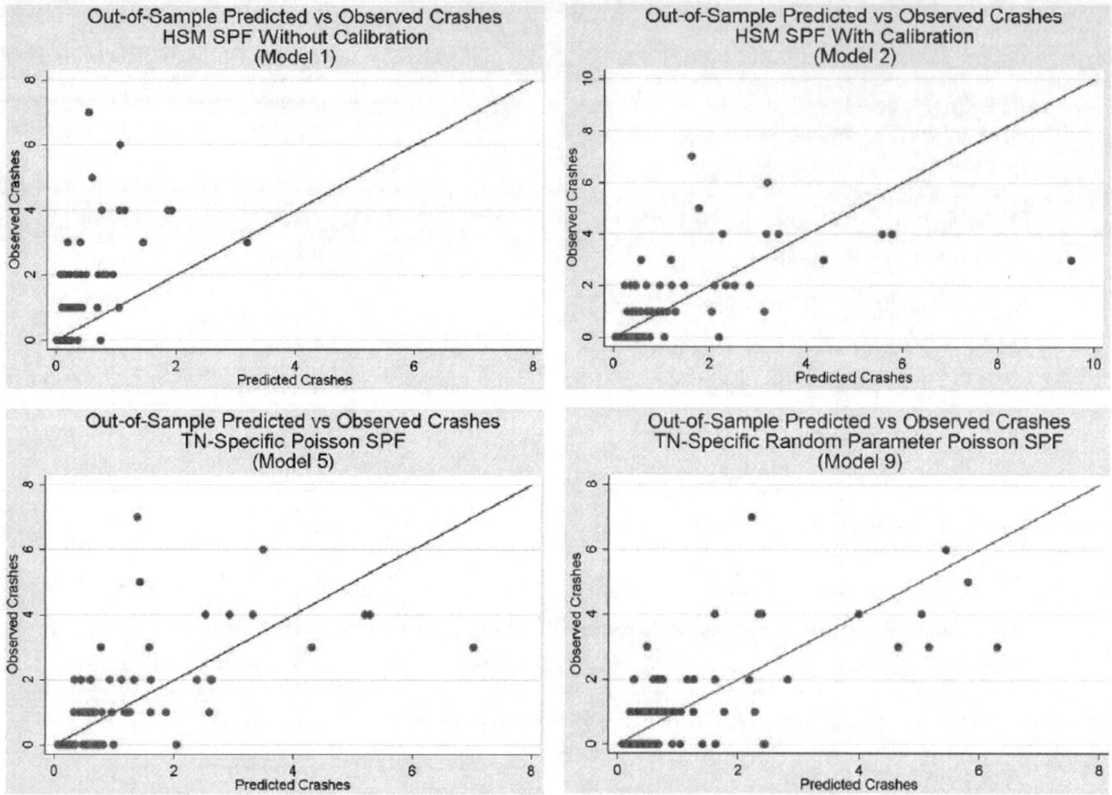

Figure 2. Out-of-sample predicted vs. observed crashes for HSM SPF (with and without calibration) and TN-specific SPFs.
Note: Based on fixed and random parameter Poisson SPFs (Red line indicates the equivalence of mean-estimated and mean-observed values).

advantage of random parameter SPFs is that more realistic predictions of crashes can be obtained that can better identify potential for safety improvement (PSI) sites, and appropriate countermeasures can be developed. To facilitate application of aforementioned SPFs in engineering practice, graphical user interfaces are needed.

Conclusions

The main objectives of this study are to apply HSM predictive models for rural two-lane, two-way roads, compute calibration factors, and explore the need for developing Tennessee-specific SPFs. Before conducting detailed empirical analysis of rural two-lane, two-way road safety, crash rates are analyzed as an effective "*first brush*" tool to quantify relative safety in different regions within Tennessee. For development of SPFs, several functional forms based on HSM guidelines and other prevalent in the literature are considered. In particular, unobserved heterogeneity is accounted for by developing fixed- and random-parameter count data models that can be used for more accurate prediction of crashes on Tennessee rural two-lane, two-way roads.

The average crash / 100-million VMT for all regions is 287.62 with a standard deviation of 384.19. Based on region-wide crash rate analysis, Region 4 (Memphis area) appears to have the lowest crash rate at 239.34 crashes /100-million VMT, followed by Region 3 (Nashville), with 282.93 crashes /10-million VMT. The calibration of HSM SPFs for rural two-lane, two-way roads revealed that average 5-year calibration factor for all regions is 2.5. This shows that these road segment crashes are at least 1.5 times greater than what HSM SPF predicts after applying calibration factors. However, some differences are observed for regionwide calibration factors as opposed to collective calibration factors for all regions. For instance, whereas average 5-year calibration factors for Region 1 and 2 are around 2.5, the calibration factors for Region 3 and 4 are different from other regions. Specifically, average 5-year calibration factors for Region 3 and 4 are around 2.8 (highest among all regions) and 2.0 (lowest among all regions), respectively. Overall, calibration factor results suggest observed crashes on Tennessee rural two-lane, two-way roads are significantly greater than crashes predicted by calibrated HSM SPFs.

The calibration of HSM SPF to match local conditions of Tennessee improved the prediction accuracies of HSM SPF;

especially, out-of-sample forecast errors for calibrated HSM SPF are found to be lower than Tennessee-specific SPFs based on including all variables in model specifications and almost similar to Tennessee-specific SPF based on HSM functional forms. However, Tennessee-specific Poisson SPF based on logarithms of AADT and segment length as only covariates performed better than calibrated HSM SPF and Tennessee-specific SPF based on HSM functional form. Finally, fully specified Tennessee-specific random parameter Poisson SPF outperformed all competing SPFs in forecasting out-of-sample crashes on Tennessee rural two-lane, two-way roads. This provides compelling empirical evidence that unobserved heterogeneity should be accounted for in estimating Tennessee-specific SPFs, and ignoring this can result in biased and inaccurate inferences and forecasts. The methodology provided in this study offers a way to quantify heterogeneous correlations of safety performance (e.g. a relationship between AADT and crash frequency) and identify the PSI for each site. Appropriate countermeasures can be developed for sites with larger magnitudes of parameters, which imply greater PSIs. While the present study accounts for random (unobserved) heterogeneity, systematic heterogeneity (e.g. because of non-linearity) is not comprehensively covered. As future work, both systematic and random heterogeneity components are recommended to be further explored in crash frequency modeling. Without explicitly accounting for both systematic and random heterogeneity, it is impossible to discern the true source of heterogeneity (systematic or random).

Acknowledgments

The authors would like to thank Tennessee Department of Transportation for supporting this research. Special thanks are due to Mr. Steve Allen, Mr. Jeff Murphy, Mr. Zane Pannell, Mr. David A. Duncan for their timely guidance in data collection efforts.

Support was provided by the U.S. DOT funded Southeastern Transportation Center and the U.S. DOT through the Collaborative Sciences Center for Road Safety, a consortium led by The University of North Carolina at Chapel Hill in partnership with The University of Tennessee. The contribution of Ms. Megan Lamon in proof-reading the manuscript is highly appreciated. The first author dedicates his part of the research effort to his dearest mother.

References

1. American Association of State Highway Transportation Officials (AASHTO). *Highway Safety Manual* (Vol. 2). Washington, D.C., 2010.
2. Sun, X., Y. Li, D. Magri, and H. Shirazi. Application of Highway Safety Manual Draft Chapter: Louisiana Experience. *Transportation Research Record: Journal of the Transportation Research Board*, 2006. 1950: 55–64.
3. Brimley, B., M. Saito, and G. Schultz. Calibration of Highway Safety Manual Safety Performance Function: Development of New Models for Rural Two-lane Two-way Highways. *Transportation Research Record: Journal of the Transportation Research Board*, 2012. 2279: 82–89.
4. Mehta, G., and Y. Lou. Calibration and Development of Safety Performance Functions for Alabama: Two-lane, Two-way Rural Roads and Four-lane Divided Highways. *Transportation Research Record: Journal of the Transportation Research Board*, 2013. 2398: 75–82.
5. Mannering, F. L., and C. R. Bhat. Analytic Methods in Accident Research: Methodological Frontier and Future Directions. *Analytic Methods in Accident Research*, Vol. 1, 2014, pp. 1–22.
6. Akbar, M., R. Khan, M. T. Khan, B. Alam, M. Elahi, B. Wali, and A. A. Shah. Methodology for Simulating Heterogeneous Traffic Flow at Intercity Roads in Developing Countries: A Case Study of University Road in Peshawar. *Arabian Journal for Science and Engineering*, Vol. 43, No. 4, 2017, pp. 2021–2036.
7. Arvin, R., M. Khademi, and H. Razi-Ardakani. Study on Mobile Phone Use While Driving in a Sample of Iranian Drivers. *International Journal of Injury Control and Safety Promotion*, Vol. 24, No. 2, 2017, pp. 256–262.
8. Liu, J., A. J. Khattak, and B. Wali. Do Safety Performance Functions Used for Predicting Crash Frequency Vary Across Space? Applying Geographically Weighted Regressions to Account for Spatial Heterogeneity. *Accident Analysis & Prevention*, Vol. 109, 2017, pp. 132–142.
9. Khattak, A. J., and B. Wali. Analysis of Volatility in Driving Regimes Extracted from Basic Safety Messages Transmitted Between Connected Vehicles. *Transportation Research Part C: Emerging Technologies*, Vol. 84, 2017, pp. 48–73.
10. Lord, D., and F. Mannering. The Statistical Analysis of Crash-frequency Data: A Review and Assessment of Methodological Alternatives. *Transportation Research Part A: Policy and Practice*, Vol. 44, No. 5, 2010, pp. 291–305.
11. Kamrani, M., B. Wali, and A. J. Khattak. Can Data Generated by Connected Vehicles Enhance Safety? Proactive Approach to Intersection Safety Management. *Transportation Research Record: Journal of the Transportation Research Board*, 2017. 2659: 80–90.
12. Washington, S. P., M. G. Karlaftis, and F. L. Mannering. *Statistical and Econometric Methods for Transportation Data Analysis*, 2nd ed. CRC press, Florida, 2010.
13. Khattak, A. J., A. J. Khattak, and F. M. Council. Effects of Work Zone Presence on Injury and Non-injury Crashes. *Accident Analysis & Prevention*, Vol. 34, No. 1, 2002, pp. 19–29.
14. Li, X., A. J. Khattak, and B. Wali. Large-Scale Traffic Incident Duration Analysis: The Role of Multi-agency Response and On-Scene Times. *Transportation Research Record: Journal of the Transportation Research Board*, 2017. 2616: 39–48.
15. Wali, B., A. Ahmed, and N. Ahmad. An Ordered-Probit Analysis of Enforcement of Road Speed Limits. *Proceedings of the Institution of Civil Engineers-Transport*, Thomas Telford Ltd, Westminster, 2017, pp. 1–10.
16. Bozdogan, H. Model Selection and Akaike's Information Criterion (AIC): The General Theory and Its Analytical Extensions. *Psychometrika*, Vol. 52, No. 3, 1987, pp. 345–370.

17. E-TRIMS, Enhanced Tennessee Roadway Information Management System, Tennessee Department of Transportation electronic system for crashes and road inventory, http://tsite.org/wp-content/uploads/2012/11/E-TRIMS-TSITE-2014.pdf. Accessed April 6, 2018.
18. TDOT. Traffic History. https://www.tdot.tn.gov/APPLICATIONS/traffichistory. Accessed April 7, 2018.
19. Shmueli, G. To explain or to Predict? *Statistical Science*, Vol. 25, No. 3, 2010, pp. 289–310.
20. El-Basyouny, K., and T. Sayed. Accident Prediction Models With Random Corridor Parameters. *Accident Analysis & Prevention*, Vol. 41, No. 5, 2009, pp. 1118–1123.

The Standing Committee on Highway Safety Performance (ANB25) peer-reviewed this paper (18-00065).

The views expressed in this paper are those of the authors.

TRR
JOURNAL OF THE TRANSPORTATION RESEARCH BOARD

Article

Transportation Research Record
2018, Vol. 2672(30) 21–31
© National Academy of Sciences:
Transportation Research Board 2018
Article reuse guidelines:
sagepub.com/journals-permissions
DOI: 10.1177/0361198118768525
journals.sagepub.com/home/trr
SAGE

Evaluating the Safety Effects of Span Wire to Mast Arm Signal Conversion

Zulqarnain H. Khattak[1] and Michael D. Fontaine[2]

Abstract

The mast arm signal mounting configuration has often been assumed to provide greater conspicuity and visibility than span wire installations, but the impacts of this countermeasure on road safety have not been well established quantitatively. This paper's intent is to provide a comprehensive evaluation of safety effects of converting span wire installations to mast arm installations. An Empirical Bayes before–after safety analysis was conducted using crash data for 30 intersections collected from the Virginia Department of Transportation (VDOT). A crash type analysis was also conducted to observe changes in crash types. Based on the data available, no statistically significant improvement in safety was found. Average crash modification factor (CMF) values of 0.97 and 0.98 were observed for total and fatal and injury crashes, but these CMFs were not statistically significant. A reduction in the proportion of angle crashes was observed, but the change was not statistically significant and the overall distribution of crashes did not change significantly after converting signal configurations from span wire to mast arm. Although mast arm deployments may offer advantages in maintenance costs and aesthetics, there does not appear to be a substantial safety benefit to these conversions.

Traffic signals play a vital role in managing conflicting traffic movements at intersections. According to the American Association of State Highway and Transportation Officials (AASHTO), on average there are five crashes at intersections every minute, and one person dies every hour of every day at an intersection somewhere in the United States (*1*). Although many recent innovations in signal operations and control have been able to generate safety improvements, the basic configuration and conspicuity of traffic signals remains an important factor in signalized intersection safety. Over the years, several engineering countermeasures have been proposed to improve safety and efficiency of intersections by improving signal visibility and conspicuity (*2*).

The visibility and conspicuity of traffic signals is of prime importance because poor conspicuity may lead to drivers failing to detect the signals. The Manual on Uniform Traffic Control Devices (MUTCD) provides guidelines for standard signal installation configurations such as dimensions (size and number of faces) and position (horizontal or vertical) on the road (*3*). These standards are provided to ensure consistency in signal design throughout the United States and improve driver familiarity. However, various traffic agencies practice a wide variety of signal mounting configurations including signals mounted on poles located at corners of streets (pedestal mount), signals hung overhead on span wires attached to poles at intersections, and signals mounted on overhead mast arms. Examples of span wire and mast arm installations are shown in Figure 1.

Both overhead signal configurations (span wire and mast arm) have advantages over the pedestal mounted signals in signal visibility. Pedestal mounted signals lack the conspicuity of overhead signals, and may be more difficult to detect in a dense urban environment. Although span wire signals are more visible than pedestal mounted signals, they may be considered less aesthetically pleasing than mast arm configurations because of long overhead wires. They also may have higher maintenance costs as compared with mast arm (*4, 5*). Both mast arm and span wire installations may have issues at windy locations, but mast arms have fewer issues and also provide better support for modern ITS devices (cameras and radars) used for vehicle detection (*5*). The potential aesthetic, maintenance, and operational benefits of mast arm installations are offset by a higher initial construction cost.

Although various factors that contribute to intersection safety have been widely studied (*2*), changing signal configuration from span wire to mast arm has not been examined. It has often been assumed in Virginia that mast arm installations may be inherently safer as the mast arm is perceived as increasing the visibility and conspicuity of traffic signals. Recently, several projects have deployed mast arms as part of a set of safety countermeasures. This study assesses whether there is empirical evidence as to whether the mast arm installations created safety improvements on those projects.

[1]Department of Civil and Environmental Engineering, University of Virginia, Charlottesville, VA
[2]Virginia Transportation Research Council, Charlottesville, VA

Corresponding Author:
Address correspondence to Zulqarnain H. Khattak: zk6cq@virginia.edu

Figure 1. Examples of span wire and mast arm installations, showing images before and after conversion at a site: (*a*) span wire installation (before conversion), and (*b*) mast arm installation (same site, after conversion).

Literature Review

Several studies have evaluated the impact of signal visibility and conspicuity on safety. El-Basyouny and Sayed evaluated the safety effectiveness of using larger signal heads and reflective backplates at nine intersections using a before and after full Bayes approach with comparison groups (*6*). The results revealed a reduction in crashes, with a crash modification factor (CMF) of 0.71 for fatal and injury crashes and 0.79 for property damage only (PDO) crashes. Miska et al. evaluated the safety impact of improving traffic signal conspicuity by adding yellow prismatic reflective tape along the edges of signal backplates (*7*). Auto insurance claims data for 25 locations in British Columbia were used in this evaluation. The treatment effectiveness was evaluated as an odds ratio for the number of claims within treatment and comparison groups. The results revealed reductions in total claims at 19 sites after installation of the reflective backplates.

Relatively little literature is available on the safety benefits of signal mounting configuration (*8–11*). Schattler et al. evaluated 12 signalized intersections, consisting of five mast arm sites, five pedestal mounted sites, and two span wire sites (*8*). Sites were evaluated based on both safety and efficiency. A total of 252 observations were recorded for red light running within a period of 3 h during weekdays. Lower red light running rates were observed for mast arm sites as compared with span wire and pedestal mounted sites. Wundersitz analyzed the existing literature on traffic signal conspicuity and road safety (*9*). The report reviewed literature describing crash reductions observed by installing additional signal heads at mast arm sites and conversion of signal installations from pedestal to mast arm. Their review found that the use of mast arms is associated with varying benefits ranging from an 89% reduction to a 29% increase in total crashes. The greatest reduction was observed in right-angle crashes. The study concluded by stating that these findings were not conclusive because many of the evaluations reviewed suffered from methodological limitations, such as the use of simple before and after studies and failure to account for traffic flow and regression to the mean (RTM) effects. Bhesania studied the safety impact of replacing pedestal mounted signal heads with mast arm mounted signal heads while simultaneously adding all-red clearance intervals at six intersections in Kansas City, Missouri (*10*). Results based on a simple before–after study showed that upgrading to mast arm signals on wide streets and implementing an all-red interval are cost-effective improvements to reduce the number of collisions at intersections. The upgrade was found to be effective for intersections with pedestal mounted signals that experienced a larger number of right-angle collisions, resulting in an overall 25% reduction in total crashes and a 63% reduction in right-angle crashes. These reductions are likely to be overly optimistic, however, given the limitations of the study. The naïve before and after study design, confounding of mounting configuration with all-red clearance, and lack of correction for RTM bias create some concerns about the reproducibility of these results.

Several studies were able to document safety improvements from converting pedestal mounted signals to mast arm configurations specifically. Thomas and Smith analyzed 31

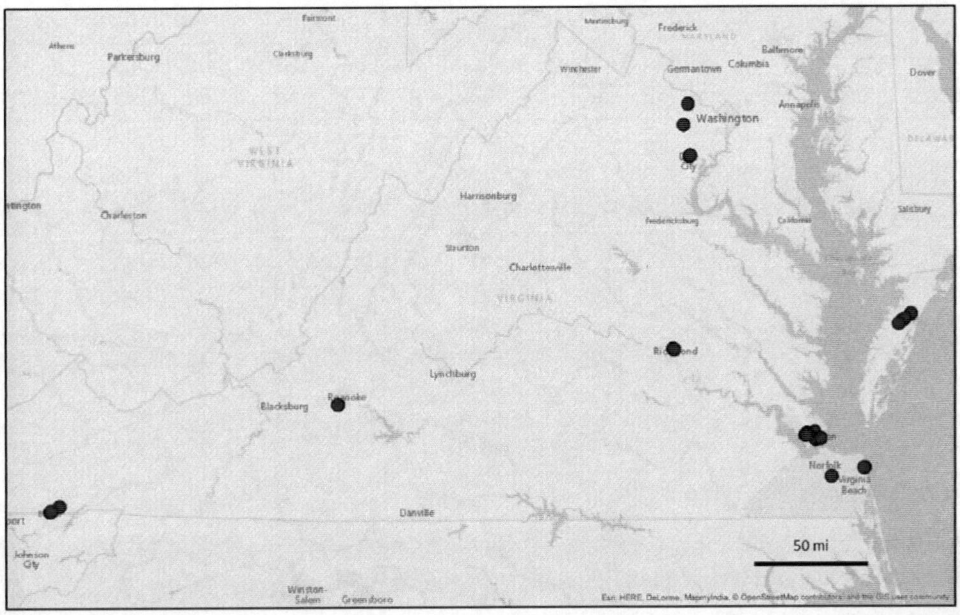

Figure 2. Map of study intersections.

Iowa intersections where pedestal mounted signals were replaced with mast arm signals using a simple before and after crash comparison (*12*). The study concluded that replacing pedestal mounted signals with a mast arm configuration is likely to result in a 32% reduction in total crashes, but no statistically significant conclusions can be drawn about fatalities. Another study evaluated the safety impact of converting pedestal mounted signals to mast arms using a simple before–after crash comparison and found a reduction of 59% in all collisions, a 44% reduction in fatal and injury collisions, a 51% reduction in property damage collisions, and a 22% reduction in angle collisions (*4*).

Kwigizile et al. compared safety performance of box span installations to diagonal span installations, without making a clear distinction between span wire or mast arm mounting (*11*). The two configurations were assessed using a driver survey and crash data analysis. The survey results showed that drivers preferred diagonal span signals compared with box span because they provided better visibility and lane recognition. A before–after with comparison group analysis using limited crash data revealed that box span signal installations had 3% fewer total crashes and 11% fewer fatal and injury crashes, but those results were not statistically significant.

Objectives and Scope

Although prior studies have found significant differences in the safety effects of pedestal mounted versus mast arm configurations, no rigorous comparison of mast arm and span wire installations was identified in the literature. The objective of this research is to determine the safety effects of changing signal configuration from span wire to mast arm by developing CMFs for the conversion. This study aims to answer the following questions:

1) Are there any safety benefits associated with span wire to mast arm conversion?
2) Does crash type distribution change following the conversion of span wire to mast arm?

The safety effects of mast arm traffic signals were evaluated using regional and statewide safety performance functions (SPFs) developed for the state of Virginia. CMFs were computed using applicable SPFs to evaluate the reduction in total and fatal and injury crash categories. Knowledge on the extent to which this countermeasure may help in crash reduction could be beneficial in making future deployment decisions about mast arm traffic signals.

Data Description

The Virginia Department of Transportation (VDOT) provided site and installation date information on 78 intersections that had been converted from span wire to mast arm within the last 10 years. After examining available annual average daily traffic (AADT) data and crash data, this list was reduced to 30 intersections for which all data elements were present and at least 2 years of post-conversion data were available. The study locations are shown in Figure 2.

Table 1. Summary of Site Description

VDOT District	Intersection#	Years of before data	Years of after data	Posted speed limit (mi/h)	Before period: Average $AADT_{maj}$	Before period: Average $AADT_{min}$	Before period: ΣTotal crash	Before period: ΣFI crash	After period: Average $AADT_{maj}$	After period: Average $AADT_{min}$	After period: ΣTotal crash	After period: ΣFI crash
Northern Virginia	1	3	2	40	14718	4119	6	4	13994	4037	0	0
	2	3	2	40	33566	7617	20	5	34587	8617	19	10
	3	3	2	45	33566	5045	46	12	34062	5240	10	3
	4	3	2	45	26919	25865	72	30	26661	11862	25	10
Hampton Roads	1	3	2	35	32772	31443	27	14	30534	29154	10	8
	2	3	2	35	19621	3364	14	4	18463	3038	4	2
	3	3	2	45	13389	8389	18	6	8440	4522	6	2
	4	3	2	35	24122	4895	16	7	16407	3166	3	0
	5	3	2	35	6973	4350	12	5	6573	3663	5	2
	6	3	2	25	6567	4449	4	1	4077	3663	7	1
	7	3	2	35	7013	3145	9	3	6573	2874	9	1
	8	3	2	25	6568	3145	3	1	4032	2874	11	6
	9	3	2	35	3145	1589	4	0	2874	1685	13	6
	10	3	2	40	16723	6729	8	6	12606	5591	2	0
	11	3	2	45	12260	9274	43	17	9232	8332	15	4
	12	3	2	40	19621	4311	17	5	18463	3665	8	2
	13	3	2	55	19270	3122	24	10	17399	2783	7	1
	14	3	2	55	19270	2957	3	3	17399	2397	3	0
	15	3	2	55	19270	1229	15	8	17399	1302	8	1
	16	3	2	45	18832	1783	24	6	20758	1452	18	6
Richmond	1	3	3	25	32326	15372	29	12	31986	15583	23	6
	2	3	3	25	23045	4067	10	3	26005	3633	13	4
	3	3	3	35	23045	6458	12	2	26005	6366	11	2
	4	3	3	25	23045	4067	6	1	26005	3633	6	2
	5	3	3	25	46185	1084	12	3	15583	1193	5	4
	6	3	3	25	46185	880	8	3	15583	476	6	2
Bristol	1	3	2	35	16661	8935	24	8	13972	7834	8	4
	2	3	2	35	13111	2419	19	7	12607	2193	3	0
	3	3	2	25	19708	8582	27	4	17984	7859	22	7
Salem	1	4	5	35	29470	27055	76	22	16114	15688	51	16
Σ of total and FI crashes for all sites						Before	608	212		After	331	112

Note: AADT = annual average daily traffic; FI = fatal and injury.

Crash data for the selected intersections were collected from the VDOT Roadway Network System (RNS). A buffer of 250 ft around the intersection was used to define intersection crashes. This definition is consistent with Virginia intersection SPFs used by VDOT, discussed later in this paper. Most of the installations were constructed in 2012 and 2013. Data from the year of construction were removed from the analysis to reduce the potential influence of construction activities and initial driver adjustment to the treatment. AADT information was collected from the VDOT traffic monitoring system. The VDOT system had data for both before and after the deployment period on all approach legs. Deviations from SPF base conditions were collected by visually reviewing the intersections on Google Earth. Base conditions used were the same as the AASHTO Highway Safety Manual (HSM) (*13*) and included number of left turn lanes, right turn lanes, right turn on red, left turn phasing, and presence of left turn phasing, The descriptive statistics for all study sites are provided in Table 1 and site conditions are shown in Table 2.

Methodology

Empirical Bayes Method

The Empirical Bayes (E-B) before–after predictive method prescribed in HSM (*13*) was used to estimate the CMFs for conversion from span wire to mast arm signal configuration. The E-B method is much more reliable and rigorous than simple before–after analysis methods because it takes observed crash frequency into account and combines it with long-term expected crash frequencies computed through

Table 2. Summary of Conditions at Study Sites

No. of legs	Right turn lanes: Approaches with turn lanes	Right turn lanes: No. of sites	Left turn lanes: Approaches with turn lanes	Left turn lanes: No. of sites	No. of sites where right turn on red is permitted on at least 1 approach	No. of sites with roadway lighting	Left turn phasing: All protected	Left turn phasing: All permitted	Left turn phasing: Mix of protected and permitted
3	0	2	0	2	4	4	1	2	2
	1	3	1	2					
	2	0	2	1					
	3	0	3	0					
4	0	1	0	1	21	17	8	5	12
	1	4	1	0					
	2	12	2	11					
	3	4	3	4					
	4	4	4	9					

Note: No. = number.

SPFs. The E-B method reduces RTM bias associated with traditional crash rate and frequency-based safety evaluation methods. A recent study also compared CMFs for total crashes and run off the road crashes using E-B before–after, cross sectional, and propensity scores-potential outcomes and found that all three methods yielded similar results, with E-B providing results having lowest standard errors (*14*). This suggests that E-B should be used as a preferred method of analysis when it is both possible and practically feasible. The three basic assumptions of the E-B method are that: (1) the number of crashes at any site follows the Poisson distribution; (2) the mean of a population can be approximated by a gamma distribution; and (3) changes from year to year unaccounted for factors are similar for all reference sites (*15*). All of these assumptions and data requirements of E-B method were satisfied and the step by step procedure for E-B method used is provided in Figure 3 (*18*).

After collecting relevant data and classifying the study intersections as urban/suburban and rural, the predicted average crash frequencies for both before and after periods were estimated using SPFs, which are negative binomial regression models developed using sites with pre-defined base conditions. HSM provides national SPFs, but SPFs for the state of Virginia were previously developed in other research (*17*). Those SPFs for Virginia were utilized for estimating CMFs in this paper. Virginia SPFs were created for three regions based on variations in topography and driver population: Northern Virginia (which is very urbanized and congested), the Western region (which is mountainous and predominantly rural), and the Eastern region (which is mainly flat, and contains a mixture of urban and rural sites). Separate SPFs were developed for each region, but SPFs for some regions had higher standard errors for some combinations of intersection configurations and crash types than the statewide model (representing lower quality). In those cases, statewide SPFs were preferred for those regions. The general SPF equation is given by Equation 1, and the regional and statewide SPFs used for the three regions are shown in Table 3. In this study, only SPFs for total and fatal and injury crashes were examined. Unlike the national SPFs in the HSM, the Virginia SPFs are for all crash types and do not distinguish between single and multivehicle crashes.

$$N_{spf} = \exp^{\alpha} \times \mathrm{AADT}_{max}^{\beta 1} \times \mathrm{AADT}_{min}^{\beta 2} \quad (1)$$

Once the SPFs for specific intersections were selected, they were adjusted for any deviations from base conditions using the CMFs from HSM. Equation 2 shows how these CMFs were used to adjust SPF predictions.

$$N_{predicted(b)} = N_{spf} \times \left(\mathrm{CMF}_{1x} \times \mathrm{CMF}_{2x} \times \ldots\ldots \times \mathrm{CMF}_{yx} \right) \quad (2)$$

where N_{spf} is the SPF equation and CMF_{1x-yx} represents the HSM CMFs based on conditions present at the site. The base conditions include:

- No left and right turn lanes present;
- Right turn on red permitted on all approaches;
- No roadway lighting;
- Permitted left turn phasing present on all approaches.

Based on any deviation from these base conditions at study locations, CMFs were multiplied with the crash frequency N_{spf} obtained from SPFs. The values for CMFs were taken from HSM Tables 12–24 to 12–27 for urban intersections and from HSM Tables 10–13 to 10–14 for rural intersections. These CMFs are shown in Tables 4 and 5 and Equations 3 and 4.

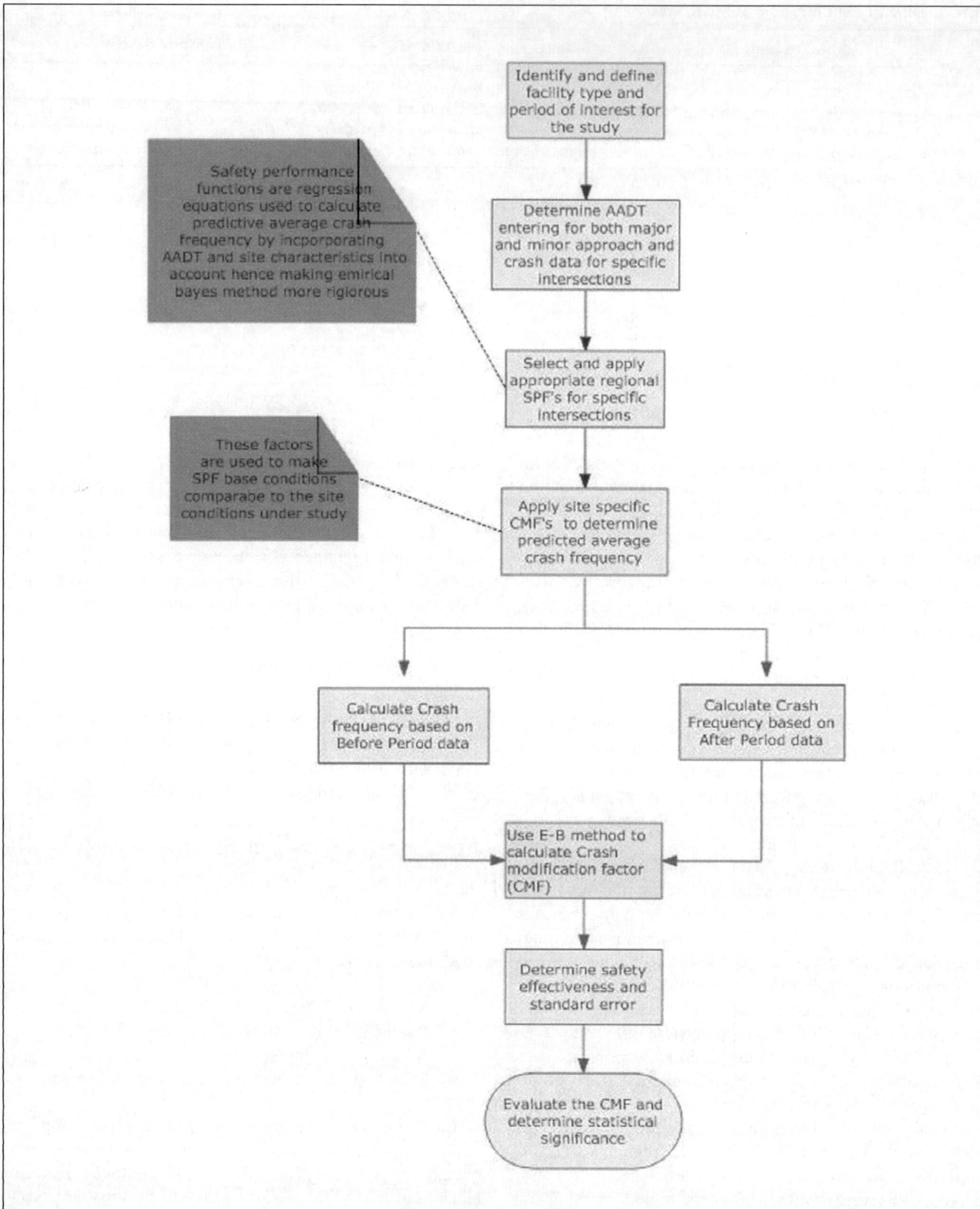

Figure 3. Flow chart for using E-B method to estimate CMF (*16,18*).

Table 3. Regional SPFs for Urban/Suburban and Rural Intersections in Virginia (*17*)

Type of intersection	Region	Crash severity	Safety performance function (SPF)	Over-dispersion parameter (*K*)
Urban 4-legged signalized	Statewide	Total	$\left(exp^{-7.6234} \times AADT_{max}{}^{0.6742} \times AADT_{min}{}^{0.3453}\right)$	0.6712
		Fatal and injury	$\left(exp^{-8.5256} \times AADT_{max}{}^{0.6477} \times AADT_{min}{}^{0.3579}\right)$	0.6296
	Western	Total	$\left(exp^{-12.3931} \times AADT_{max}{}^{0.631} \times AADT_{min}{}^{0.4567}\right)$	0.9771
		Fatal and injury	$\left(exp^{-11.4284} \times AADT_{max}{}^{0.8662} \times AADT_{min}{}^{0.4412}\right)$	0.7947
	Eastern	Total	$\left(exp^{-8.8553} \times AADT_{max}{}^{0.7825} \times AADT_{min}{}^{0.3706}\right)$	0.5727
		Fatal and injury	$\left(exp^{-9.9582} \times AADT_{max}{}^{0.7484} \times AADT_{min}{}^{0.4017}\right)$	0.5828
Urban 3-legged signalized	Statewide	Total	$\left(exp^{-6.5430} \times AADT_{max}{}^{0.6591} \times AADT_{min}{}^{0.2119}\right)$	0.7695
		Fatal and injury	$\left(exp^{-8.4268} \times AADT_{max}{}^{0.7147} \times AADT_{min}{}^{0.2481}\right)$	0.8387
	Western	Total	$\left(exp^{-9.6143} \times AADT_{max}{}^{0.8677} \times AADT_{min}{}^{0.3297}\right)$	0.8674
		Fatal and injury	$\left(exp^{-11.0104} \times AADT_{max}{}^{0.9080} \times AADT_{min}{}^{0.3226}\right)$	1.0307
Rural 4-legged signalized	Statewide	Total	$(exp^{-6.9589} \times AADT_{max}{}^{0.6746} * AADT_{min}{}^{0.253}$	0.5643
		Fatal and injury	$\left(exp^{-8.7116} \times AADT_{max}{}^{0.7839} \times AADT_{min}{}^{0.2166}\right)$	0.5553
	Western	Total	$\left(exp^{-6.3951} \times AADT_{max}{}^{0.5508} \times AADT_{min}{}^{0.3106}\right)$	0.3491
		Fatal and injury	$\left(exp^{-8.0583} \times AADT_{max}{}^{0.6809} \times AADT_{min}{}^{02557}\right)$	0.4422
	Eastern	Total	$\left(exp^{-0.6468} \times AADT_{max}{}^{0.2262} \times AADT_{min}{}^{-0.0158}\right)$	0.4329
		Fatal and injury	$\left(exp^{-4.4381} \times AADT_{max}{}^{0.4342} \times AADT_{min}{}^{0.1523}\right)$	0.4004
Rural 3-legged signalized	Statewide	Total	$exp^{-7.5754} \times AADT_{max}{}^{0.6465} \times AADT_{min}{}^{0.3332})$	0.7855
		Fatal + Injury	$\left(exp^{-8.7112} \times AADT_{max}{}^{0.7224} \times AADT_{min}{}^{0.2578}\right)$	0.7489

Table 4. CMFs for Presence of Left and Right Turn Lanes

Intersection type	Type of turn lanes	No. of legs	Number of approaches with right or left turn lanes: One	Two	Three	Four
Urban	Right turn	3	0.96	0.92	—	—
		4	0.96	0.92	0.88	0.85
	Left turn	3	0.93	0.86	0.80	—
		4	0.90	0.81	0.73	0.63
Rural	Right turn	3	0.86	0.74	—	—
		4	0.96	0.92	0.88	0.85
	Left turn	3	0.56	0.31	—	—
		4	0.82	0.67	0.55	0.45

Note: No. = number.

Table 5. CMFs for Type of Left Turn Phasing

Type of turn phasing	CMF
Permissive	1.00
Protected/permissive	0.99
Protected	0.94

The CMF for right turn on red prohibition was calculated using Equation 3, where n_{proh} stands for number of approaches where right turn on red was prohibited. The base condition is that right turn on red is permitted on all approaches of a signalized intersection.

$$CMF_{\text{right turn}} = CMF = 0.98^{n_{\text{proh}}} \tag{3}$$

Similarly, the CMF for the effect of presence of lighting was calculated using Equation 4. The variable P_{night} represents the proportion of night time crashes at each intersection, which was calculated using the number of night time crashes in the observed crash data in the before period. This value was set as 0.18 for urban and 0.23 for rural intersections. The base condition is the absence of lighting.

$$\text{CMF}_{\text{lighting}} = 1 - 0.38 \times p_{\text{night}} \quad (4)$$

After calculating the predicted average crash frequencies, the E-B correction was applied by weighting observed and predicted crash frequencies (N) to calculate an expected average crash frequency in the before period using

$$N_{\text{expected},B} = w_{i,B} \times N_{\text{predicted}} + \left(1 - w_{i,B}\right) \times N_{\text{observed},B} \quad (5)$$

where $W_{i,B}$ represents the weighted adjustment, and is calculated using

$$W_{i,B} = \frac{1}{1 + k \sum N_{\text{predicted}}} \quad (6)$$

It is worth noting that Equation (5) represents the crash estimates in terms of expected average crash frequency in the before period and the E-B method also requires crash estimates for the after deployment period. As a result, an adjustment (r_i) is made using Equation (7) to account for differences between the before and after periods of crash data in the duration, number of years, and traffic volumes.

$$r_i = \frac{\sum N_{\text{predicted},A}}{\sum N_{\text{predicted},B}} \quad (7)$$

Finally, the expected crash frequency in the after deployment period was calculated by

$$N_{\text{expected},A} = N_{\text{expected},B} \times r_i \quad (8)$$

The overall effectiveness (OR) of the treatment in terms of index of effectiveness was calculated by:

$$OR' = \frac{\sum_{\text{All sites}} N_{\text{observed},A}}{\sum_{\text{All sites}} N_{\text{expected},A}} \quad (9)$$

The CMF which indicates the safety benefit in terms of relative change in crash frequency because of a change in specific countermeasure (mast arm conversion in this case) was calculated by

$$\text{CMF} = \frac{OR'}{1 + \frac{Var(\sum_{\text{All sites}} N_{\text{expected},A})}{\left(\sum_{\text{All sites}} N_{\text{expected},A}\right)^2}} \quad (10)$$

Based on Hauer (*15*):

$$Var\left[\sum_{\text{All sites}} N_{\text{expected},A}\right] = \sum_{\text{All sites}} \left[\begin{matrix}(r_i)^2 \times N_{\text{expected},B} \\ \times (1 - w_{i,B})\end{matrix}\right] \quad (11)$$

Although CMF is a point estimate for safety benefit, the safety effectiveness in terms of percentage was calculated by

$$\text{Safety Effectiveness} = (1 - \text{CMF}) \times 100 \quad (12)$$

The standard error of overall effectiveness using information on variance of crashes is given by:

$$o' = \sqrt{\frac{(OR')^2 \left[\frac{1}{N_{\text{observed},A}} + \frac{Var(\sum_{\text{All sites}} N_{\text{expected},A})}{(\sum_{\text{All sites}} N_{\text{eexpected}A})^2}\right]}{1 + \frac{Var\left(\sum_{\text{All sites}} N_{\text{expected},A}\right)}{(\sum_{\text{Allsitess}} N_{\text{expected},A})^2}}} \quad (13)$$

The standard error was used to assess whether any changes in crash frequency after mast arm installation were significant.

Crash Type Analysis and Test of Proportions

A countermeasure may affect particular types of crashes more than others, so it is important to analyze each crash type individually. Because the numbers of years in the before and after periods were not consistent for all of the study sites, the numbers of crashes for each crash type were converted into proportions using

$$\text{Crash Proportion} = \frac{\text{Number of Crashes for Each Crash Typee}}{\text{Total Crashes}} \quad (14)$$

The estimates from Equation (14) for each crash type in the before and after period were compared to analyze how they were affected by the countermeasure. The crash proportions were also tested for their statistical significance using the 2-proportions Z-test with a null hypothesis assuming that the two proportions were equal. The test statistics used for testing the difference between the two population proportions is given by

$$Z = \frac{p_1 - p_2}{\sqrt{p'(1-p')\left(\frac{1}{n_1} + \frac{1}{n_2}\right)}} \quad (15)$$

where p_1 and p_2 are the proportion of crash types in the before and after period and n_1 and n_2 are the total number of crashes in the two periods. The variable p' stands for common proportion in both populations or the proportion of successes. Because this value is unknown, assuming

Table 6. CMFs for All, Angle, and Rear End Crashes

Crash type	No. of sites	Crash severity	CMF	SE	Safety effectiveness	Statistical significance
All	30	Total	0.97	0.069	3%	No
		Fatal and injury	0.98	0.119	2%	No
Angle	30	Total	0.95	0.086	5%	No
		Fatal and injury	0.97	0.137	3%	No
Rear end	30	Total	1.00	0.091	0%	No
		Fatal and injury	1.20	0.199	-20%	No

Note: No. of sites = number of sites; SE = standard error.

Table 7. CMFs for Intersection and Location Type

Road and intersection type	No. of sites	Crash severity	CMF	SE	Safety effectiveness	Statistical significance
Urban/suburban, 3- or 4-leg	26	Total	0.98	0.075	2%	No
		Fatal and injury	1.03	0.139	−3%	No
Rural or urban/suburban 4-Leg	25	Total	0.99	0.076	1%	No
		Fatal and injury	1.00	0.128	0%	No
Urban/suburban 4 Leg	22	Total	1.00	0.080	0%	No
		Fatal and injury	1.09	0.146	−9%	No

Note: No. of sites = number of sites; SE = standard error.

Equation (15) follows a standard normal distribution, p' can be estimated using

$$p' = \frac{Y_1 + Y_2}{n_1 + n_2} \quad (16)$$

where Y_1 and Y_2 represent the number of crashes for each crash type in the before and after period.

Results and Discussion

E-B Analysis

The countermeasure locations selected were predominantly composed of urban/suburban intersections (26 of 30 sites), and most sites had four legs (25 of 30 sites). The data was first evaluated as an aggregate system using all sites (Table 6), and was later subdivided based on location and configuration (Table 7) using configurations in which at least 20 sites were available.

Table 6 shows the aggregate safety effects of span wire to mast arm conversion across both urban/suburban and rural intersections. A CMF value of 0.97 was determined for total crashes, indicating a reduction of 3% in total crashes. Similarly, fatal and injury crashes had a CMF of 0.98, which indicates a crash reduction of 2%. Therefore, a slight net benefit was observed for both total and fatal and injury crash categories. These reductions were not statistically significant, however. Angle and rear end crashes also exhibited small but statistically insignificant changes in crashes.

Table 7 shows the results when only homogenous categories with at least 20 sites are analyzed. Table 7 reveals that urban/suburban intersections produce similar safety benefits to those for the combined data set, with statistically insignificant CMFs. These results are not unexpected given the prevalence of urban sites in the data set. The results are again not statistically significant at 95% confidence level. Similarly, the CMFs for total and fatal and injury crash categories for 4-leg intersections are generally consistent with combined data set results. Only five 3-leg intersections were present in the data set, so insufficient data were available to calculate a separate CMF for that configuration. The CMFs for both total and fatal and injury crashes for urban 4-legged intersections also overlap with CMFs for other categories and are statistically insignificant. This analysis confirms that converting signals from span wire to mast arm configuration does not produce any significant reduction in crashes, at least for urban/suburban 4-leg intersections.

Crash Type Proportions

The earlier analysis showed that changing traffic signal configuration from span wire to mast arm did not create statistically significant changes in safety, but analysis by crash type may have been hampered by small sample sizes. As a result, a crash type analysis was performed to see if any significant difference in certain type of crash proportions could be observed. These results are shown in Tables 8 and 9.

Table 8 shows that the proportion of angle crashes decreased with the conversion of traffic signals from span wire to mast arm for both total (3.08%) and fatal and injury (8.87%) crash

Table 8. Crash Proportions for All Crash Types and All Sites

Crash type	Total crashes		Fatal and injury crashes	
	Before	After	Before	After
Rear end	248 (40.78%)	139 (41.99%)	71 (33.49%)	39 (34.82%)
Angle	241(39.63%)	121 (36.55%)	104 (49.05%)	45 (40.17%)
Side-swipe	54 (8.88%)	30 (9.06%)	9 (4.24%)	8 (7.14%)
Head on	16 (2.63%)	8 (2.04%)	9 (4.24%)	4 (3.57%)
Other	49 (8.05%)	33 (9.06%)	19 (8.96%)	16 (14.28%)
Total	608 (100%)	331(100%)	212 (100%)	112 (100%)

Table 9. 2-Proportions Z-test for All Crash Types and All Sites

Crash type	Change in total crash proportions (before–after)	*p*-Value	Change in fatal and injury crash proportion (before–after)	*p*-Value
Rear end	−1.20%	0.72	−1.23%	0.81
Angle	3.08%	0.35	8.87%	0.12
Side-swipe	−0.18%	0.92	−2.90%	0.30
Head on	0.60%	0.84	0.67%	0.76
Other	−1.91%	0.33	−5.32%	0.17

Note: Before = span wire installation (before conversion); After = mast arm installation (same site, after conversion).

categories. Although a decrease in angle crashes was expected because of better visibility with the mast arm traffic signals, this decrease in crash proportions was offset by increases in all other type of crashes. These changes in crash proportions were also analyzed for their statistical significance based on a Z-test of proportionality to discover whether any statistically significant difference existed between the two crash periods. The results for the proportionality test are provided in Table 9. Table 9 reveals that neither the reduction in angle and head on collision proportions nor the increase in rear end and side-swipe crash proportions is statistically significant.

The crash type analysis revealed some reductions in crash proportions that accrued evenly across all type of crashes at intersections. Along with non-significant reductions, non-significant increases were also observed in certain crash types. This confirms that this countermeasure may produce a very slight reduction in the proportion of some crashes that may be offset by the increase in certain other crash types. Thus, no significant net safety benefits are associated with this countermeasure.

Conclusions and Recommendations

This paper conducted a safety evaluation of converting traffic signals from span wire to mast arm using an E-B before–after and crash type analysis. The safety benefits of span wire to mast arm conversion have been evaluated because a rigorous safety evaluation of this conversion currently does not exist.

No statistically significant reduction in crashes because of span wire to mast arm conversion was detected in this study, at least for urban/suburban four-leg intersections. An aggregate CMF value of 0.97 and 0.98 for total and fatal and injury crashes was determined, whereas CMFs of 0.95 and 0.97 were observed for total and fatal and injury crash categories for angle crashes. Because the CMFs are not statistically significant, these results imply no significant difference from CMF of 1 (0% reduction in crashes). Although there may be economic, maintenance, or aesthetic reasons to use mast arms, the data analyzed does not show conclusive safety benefits over span wire installations at urban four-leg intersections. These findings will provide planners and researchers with a tool for evaluating the net gain of this conversion before considering it for implementation. However, further research to examine safety at rural sites or at three-leg intersections is needed to determine if the CMF results are broadly transferable.

References

1. AASHTO. Strategic Highway Safety Plan. *American Association of State Highway and Transportation Officials*, Vol. 5800, No. 202, 2005, pp. 1–38.
2. Chen, L., C. Chen, R. Ewing, C. E. Mcknight, R. Srinivasan, and M. Roe. Safety Countermeasures and Crash Reduction in New York City—Experience and Lessons Learned. *Accident Analysis and Prevention*, Vol. 50, 2013, pp. 312–322. https://doi.org/10.1016/j.aap.2012.05.009.
3. FHWA. *Manual on Uniform Traffic Control Devices for Streets and Highways*. Federal Highway Administration, Washington D.C., 2012.
4. Rodegerdts, L. A., B. Nevers, B. Robinson, J. Ringert, P. Koonce, J. Bansen, T. Nguyen, J. McGill, D. Stewart, J. Suggett, T. Neuman, N. Antonucci, K. Hardy, and K. Courage.

Signalized Intersections-Informational Guide. Publication No. FHWA-HRT-04-091. U.S. Department of Transportation, Federal Highway Administration, Washington, D.C., 2003.
5. McGee, H. W. *Making Intersections Safer: A Toolbox of Engineering Countermeasures to Reduce Red Light Running*. FHWA and Institute of Transportation Engineers, FHWA/TX-03/4027-2, Texas Transportation Institute, Tex., 2003.
6. El-Basyouny, K., and T. Sayed. A Full Bayes Multivariate Intervention Model with Random Parameters among Matched Pairs for Before–After Safety Evaluation. *Accident Analysis and Prevention*, Vol. 43, No. 1, 2011, pp. 87–94. https://doi.org/10.1016/j.aap.2010.07.015.
7. Miska, E., P. de Leur, and T. Sayed. Road Safety Performance Associated with Improved Traffic Signal Design and Increased Signal Conspicuity. *Institute of Transportation Engineers Annual Meering and Exhibit*, Philadelphia, Pennsylvania, 2002, pp. 1–11.
8. Schattler, K. L., D. McAvoy, M. T. Christ, and C. M. Glauber. Impact of Signal Mounting Configurations on Red Light Running at Urban Signalized Intersections. *Institute of Transportation Engineers*, Vol. 81, No. 2, 2011, p. 22.
9. Wundersitz, L. *An Assessment of Conspicuous Traffic Signals: Mast Arms*. Centre for Automotive Safety Research, University of Adelaide, Australia, 2009.
10.Bhesania, R. Impact of Mast-Mounted Signal Heads on Accident Reduction. *Institute of Transportation Engineers*, Vol. 61, No. 10, 1991, pp. 25–29.
11. Kwigizile, V., J. Oh, R. Van Houten, D. Prieto, R. Boateng, L. Rodriguez, A. Ceifetz, J. Yassin, J. Bagdade, and P. Andridge. *Evaluation of Michigan's Engineering Improvements for Older Drivers*. No. RC 1636. Transportation Research Center for Livable Communities, Western Michigan University, Mich., 2015.
12. Thomas, G. B., and D. J. Smith. *Effectiveness of Roadway Safety Improvements*. Final Report. 00-61. Iowa Department of Transportation, Iowa, 2001.
13. AASHTO. *Highway Safety Manual*. American Association of State Highway and Transportation Officials, Washington D.C. 2010.
14. Wood, J. S., E. T. Donnell, and R. J. Porter. Comparison of Safety Effect Estimates Obtained from Empirical Bayes Before–After Study, Propensity Scores-Potential Outcomes Framework, and Regression Model with Cross-Sectional Data. *Accident Analysis and Prevention*, Vol. 75, 2015, pp. 144–154. https://doi.org/10.1016/j.aap.2014.11.019.
15. Hauer, E. *Observational Before–After Studies in Road Safety*. Pergamon Publications, Elsevier, London, 1997.
16. Khattak, Z. H. *Evaluating the Operational and Safety Aspects of Adaptive Traffic Control Systems in Pennsylvania*. Master's thesis, University of Pittsburgh, Pittsburgh, Pennsylvania, 2016.
17. Garber, N. J., and G. Rivera. *Safety Performance Functions for Intersections on Highways Maintained by the Virginia Department of Transportation*. VTRC 11-CR1. Virginia Transportation Research Council, Charlottesville, Va., 2010.
18. Khattak, Z. H., M.J. Magalotti., and Fontaine, M. D. Estimating safety effects of adaptive signal control technology using the Empirical Bayes method. *Journal of Safety Research*, Vol. 64, 2018. https://doi.org/10.1016/j.jsr.2017.12.016.

The Standing Committee on Highway Safety Performance (ANB25) peer-reviewed this paper (18-01498).

Article

TRR
JOURNAL OF THE TRANSPORTATION RESEARCH BOARD

Transportation Research Record
2018, Vol. 2672(30) 32–42
© National Academy of Sciences:
Transportation Research Board 2018
Article reuse guidelines:
sagepub.com/journals-permissions
DOI: 10.1177/0361198118768526
journals.sagepub.com/home/trr
SAGE

Missouri-Specific Crash Prediction Model for Signalized Intersections

Boris Claros[1], Carlos Sun[1], and Praveen Edara[1]

Abstract

The Highway Safety Manual (HSM) provides guidance and tools to conduct quantitative safety analysis. Crash prediction models are used to estimate the expected number of crashes per year, by facility type, severity, and crash type. There are two approaches for applying the HSM crash prediction methodology to local conditions: (1) calibration of models provided in the HSM; or (2) development of jurisdiction-specific models. There are some instances in which model calibration may not be appropriate. To illustrate this case, 601 urban signalized four-leg intersections (U4SG) in Missouri were used to obtain the calibration factor, assess the quality of the calibration factor, and develop jurisdiction-specific models. For U4SG total crashes, the calibration factor for Missouri conditions was 3.98 (standard deviation, 0.13). The assessment of the calibration factor showed a disproportional difference between the observed data in Missouri and the HSM model. Thus, the calibration was deemed inappropriate and the development of Missouri-specific models was supported. The models were developed for severities Fatal and Injury (FI) and Property Damage Only (PDO) crashes. The predictor variables considered were intersection AADT, posted speed limit, signal control type, exclusive left turn lanes, exclusive right turn lanes, right turn on red prohibited, and facilities of interest within 1,000 ft from the intersection (bus stops, schools, and alcohol sale establishments). Functional forms for all predictor variables were optimized. The log-likelihood, inverse overdispersion, and Cumulative Residuals (CURE) plots showed satisfactory measures of model accuracy.

The Highway Safety Manual (HSM) (*1*) provides guidance and tools to conduct quantitative safety analysis with performance measures such as traffic operations, geometry, and facility type. The HSM provides a prediction methodology composed of crash prediction models that estimate the expected number of crashes per year, by facility type, severity, and crash type. The prediction methodology includes Safety Performance Functions (SPF), Crash Modification Factors (CMF), and Calibration factors (C). Equation 1 shows the general form of the HSM prediction method.

$$N_{pred} = N_{spf} \times C \times \left(CMF_1 \times CMF_2 \times \ldots \times CMF_z\right) \qquad (1)$$

where:

N_{pred} = predicted crash frequency (crashes/year);
N_{spf} = predicted crash frequency for site type SPF (crashes/year);
C = calibration factor to local conditions;
CMF_z = crash modification factor specific to a site type characteristic z.

There are two approaches for applying the HSM crash prediction methodology to local conditions: 1) calibration of models provided in the HSM, or 2) development of jurisdiction-specific crash prediction models. There is general guidance and requirements for each approach (*1–3*). Srinivasan et al. (*2*) recommended that jurisdictions should first calibrate the HSM model and assess the quality of the calibration. The criteria to assess the quality of the calibration consist of the evaluation of the magnitude of the calibration factor (i.e., much less or greater than 1.0). Using regression modeling, the implicit assumption that the observed and predicted crashes have a straight linear relationship should be verified (calibration function). If the assessment of the calibration is deemed inappropriate, the development of jurisdiction-specific crash prediction models is recommended. The development of jurisdiction-specific models is likely to enhance the reliability of the crash estimates (*1, 2*).

It appears a straightforward procedure based on the assessment of the calibration. However, there are other factors that need to be taken into account such as data requirements, statistical knowledge, priorities, and effort. With several states' experiences in developing calibration factors and jurisdiction-specific models, there is a wide range of results. The diverse data sources from different jurisdictions,

[1]Civil Engineering, University of Wisconsin – Madison, Madison, WI

Corresponding Author:
Address correspondence to Boris Claros: claros@wisc.edu

availability, accuracy, and consistency have provided insight into the implementation challenges of the HSM prediction methodology into local conditions.

This paper focuses on providing a comprehensive step-by-step case study of the implementation process of the HSM prediction methodology for urban signalized four-leg intersections (U4SG) to Missouri conditions. A thorough literature review was conducted to document the results of jurisdictions with the implementation of the HSM including calibration and model development. Using 601 U4SG intersections, the calibration factor for total crashes was developed. The quality of the calibration factor was assessed through statistical testing, Cumulative Residuals (CURE) plots as a function of the intersection AADT, and analysis of the implicit straight linear relationship between the observed and predicted crashes through Negative Binomial regression modeling. The results of the quality of the calibration factor showed that HSM prediction models for U4SG were inappropriate for Missouri conditions. Consequently, Missouri-specific crash prediction models were developed. The predictor variables included in the model were the major and minor road AADT, minor and major road posted speed limit, left turn signal control type, exclusive left turn lanes, exclusive right turn lanes, prohibited right turn movement, and other relevant facilities within 1,000 Ft. of the intersections (bus stops, alcohol sale establishments, and educational facilities). The functional forms for the predictor variables were optimized. The logistic functional form was found appropriate for the intersection AADT. The posted speed limit was found to have a significant contribution to the prediction models. It is worth mentioning that the HSM models do not explicitly include the speed limit as predictor variable for urban signalized intersection crashes.

Literature Review

This section covers exiting literature on signalized intersections sampling, data collection, crash assignment, and the results of the HSM implementation in different jurisdictions.

Sampling and Data Collection

Sites should be selected randomly from a pool of sites. For calibration, a sample size of 30–50 sites with at least 100 crashes per year is recommended (*1*). However, several studies have argued that the recommendation is not statistically supported because the sample sizes are too small to estimate reliable calibration factors for different facility types (*4–8*). Alternative approaches have been proposed in which the minimum sample size is determined based on the characteristics of the observed crashes and the expected reliability of the calibration factor—the probability that the calibration factor lies within 10% of the true calibration factor. The requirements to developed jurisdiction-specific prediction models are more demanding: 100–200 intersections with at least 300 crashes per year for the total group. A minimum of three years of data is recommended (*1–3*). Some jurisdictions may not be able to gather enough samples to develop reliable calibration factors or their own prediction models.

Sampling and data collection by facility type are very complex tasks. For instance, for signalized intersections, jurisdictions' roadway facility inventories require indicators of the number of approach legs, signal control, or designation area (urban/rural). Usually, jurisdictions keep a GIS based inventory of their roadway network. Roadways are identified as links and intersections as nodes. Each link and node has its own attributes (i.e., identification number, name, direction). Although this process may facilitate the identification of candidate sites, samples need to be visually verified—video logs, Google Earth aerial and street view images (*9–11*). Unrecorded changes to the intersection may take place during the period of analysis that may significantly influence crash occurrence (i.e., change in signal control type, additional lanes, work zone). With the link and node representation, intersections may have more than one node (i.e., divided roads, exclusive right turn lanes) which may cause the omission of several crashes if the queries are not conducted properly (*12–15*). Traffic data may not be readily available for all approach legs to an intersection, so major or parallel road AADT may be used to estimate the minor road AADT at an intersection (*9, 16*).

Intersection Crash Assignment

Crash reporting is inconsistent among jurisdictions. There are different definitions of severity levels and facility assignment. These differences may significantly influence the calibration of HSM models. The severity levels used in most prediction models in the HSM are fatal and injury (FI), Property Damage Only (PDO), and Total (TOT) crashes (*1*). Severity PDO has been found to be very inconsistent across jurisdictions since monetary threshold values are considered. Some jurisdictions rely on PDO crashes self-reporting, the PDO monetary value threshold may change (*17*), or all crashes have to be reported despite monetary value or injury type (Pennsylvania and Colorado).

Crash assignment is also inconsistent across jurisdictions. In common practice, crashes are assigned to intersections when they occur within a default radius from the intersection, at the intersection, or are intersection related. An intersection related crash is (*18*): "one which (a) the first harmful event in the accident occurs on an approach to, or exit from, an intersection, and (b) the first harmful event resulted from some activity, behavior, or control related to the movement of traffic units through the intersection." Wang et al. (*19*) found that the commonly used 250 Ft. threshold was significantly different than intersection related crashes with varied boundaries. A threshold might be used when crash data do not include facility assignment information. Savolainen et al. (*14*) selected a threshold of 0.04 miles (211 Ft.) at signalized

Table 1. Available U4SG Calibration Factors by Jurisdiction

State	Period of analysis	Number of sites	Crashes per year[a]	Severity	Crash type[b]	Calibration factor	Crash assignment criteria[c]
Florida (*12*)	2005–2009	121	737	FI	All	1.88	Intersection Related
Maryland (*13*)	2008–2010	960	7,271	TOT	All	0.48	250 ft.
Michigan (*14*)	2008–2012	1,710	13,303	TOT	SV	0.98	0.04 mi (211 ft.)
					MV	1.09	
Minnesota (29)	2009–2011	31	392	TOT	All	1.22	33 ft. or Intersection related
Missouri (*30*)	2009–2011	35	1,347	TOT	All	4.91	132 ft or intersection related
North Carolina (*31, 32*)	2007–2009	122	977	TOT	All	2.79	250 ft
	2010–2015	102	1,030	TOT	All	3.07	
Ohio (*18*)	2009–2011	883	14,111	TOT	All	3.71	Intersection related
Oregon (*33*)	2004–2006	57	230	TOT	All	1.10	250 ft or intersection related

Note: FI = fatal and injury; TOT = total crashes; MV = multiple vehicle crashes; SV = single vehicle.
[a]Crashes per year of all calibration sites.
[b]All crash types.
[c]Intersection influence area threshold or intersection related criteria.

intersections for the calibration of the HSM. Similarly, Green et al. (*20*) used a threshold of 100 Ft.

Transferability of HSM Models

The HSM models were developed using data from a couple of states and a specific period of analysis. For instance, the HSM urban signalized intersection models were developed with data from Minnesota (1998–2002) and North Carolina (1997–2003) (*21*). Persaud et al. (*22*) found mixed results with the transferability of models from other jurisdictions. Sacci et al. (*23*) evaluated transferability of the HSM methodology to Italian two-lane undivided rural roads by comparing the calibrated predictions with the local data estimates. The results showed that the base SPFs differ significantly with increasing exposure, and the CMFs revealed some bias to the site characteristics.

Additionally, base SPF models follow the Negative Binomial regression with a log-linear relationship between crash frequency and site type characteristics (*1, 2*). Although the log-linear functional form may be used for convenience and consistency purposes, it may not be the most appropriate functional form for every predictor variable, site type, and severity. As a result, statistical modeling may seem disconnected from the observed data that it is trying to describe (*24, 25*). Several publications (*24–27*) provide evidence of the importance of custom functional forms for each predictor variable and whether predictor variables should take multiplicative or additive forms based on what is being represented. The choice of functional form should be such that it resembles the form suggested in the data including peaks, valleys, and points of inflection (*27*). Developing jurisdiction-specific models instead of calibrating existing models is likely to increase the accuracy and reliability of the predictions (*2, 22, 28*).

Available Calibration Factors

Table 1 provides a summary of existing calibration factors by jurisdiction for U4SG intersections. The periods of analysis considered were between 3 and 6 years. The sample sizes were from 31 to 1,710 intersections with over 100 crashes per year. The calibration factor for Maryland was developed for FI crashes only because PDO crashes were underreported and inconsistently categorized (*13*). In Michigan, the calibration was disaggregated to focus on single vehicle (SV) and multiple vehicle (MV) crashes (*14*). The calibration factors by state showed a wide range of values. Maryland had the lowest calibration factor of 0.48. The calibration factors of Michigan and Oregon were close to the value of 1. Minnesota and Florida had calibration factors higher than 1 (1.22 and 1.88, respectively). The states of Missouri, North Carolina, and Ohio had calibration factors greater than 3. The crash assignments by jurisdiction varied with different thresholds (33, 132, 211, and 250 Ft.), intersection related criteria, or both.

Available Jurisdiction-Specific Models

Several states have developed jurisdiction-specific models for different signalized intersection categories (*9, 14, 20, 34–36*). The categories modeled included the number of lanes, divided/undivided approach legs, urban, or all signalized intersections. The periods of analysis considered were between 4 and 6 years. The number of sites used for modeling was from 46 up to 4,311 intersections. Based on the type of signalized intersection modeled there were, on average, between 3 and 26 crashes per year per intersection. The severity levels were Fatal, Fatal and Injury, Type A Injury, Type B Injury, and Total crashes. The crash types were all, single vehicle, multiple vehicle, vehicle-pedestrian, and vehicle-bicycle crashes. The prediction models for Michigan were developed following similar criteria to the HSM models—similar severities, crash types, predictor variables, functional forms, and CMFs. Pedestrian and bicycle volume counts were not collected, the AADT was used instead (*14*). All other states (Colorado, Illinois, Kentucky, Oregon, and Pennsylvania), developed jurisdiction-specific SPFs only—CMFs were not developed (*9, 20, 34–36*). The functional forms of all available models follow the commonly used power function. The inverse overdispersion for the different jurisdiction-specific models ranged from 1.00 to 17.67. Traffic operations were between 6,083 and 19,005 vpd over the minor road and from 11,867 up to 45,729 vpd over the major road. The state of Colorado (*34*) used the intersection related criteria crash assignment while all other states used different thresholds (100, 211, 250, and 300 ft) (*9, 20, 35, 36*).

Methodology

This section provides information on sampling, data collection, calibration of HSM models, and development of Missouri-specific models for urban signalized four-leg intersections (U4SG).

Sampling

A pool of 4,710 candidate U4SG intersections was generated from the MoDOT database. Each candidate intersection was carefully reviewed and a total of 601 intersections were selected. The period of analysis for calibration consisted of three years between 2012 and 2014. The period of analysis for model development consisted of five years between 2010 and 2014. The selection process consisted of reviewing intersection geometry, operations, signal control, and possible changes over the period of analysis. The exhaustive review of each selected intersection guaranteed a sample set with consistent and similar features.

Data Collection

Missouri DOT keeps an updated Transportation Management Systems (TMS) which provides detailed information of travelways, safety, traffic, and pavement. Some of the tools of MoDOT TMS include accident browser (with access to crash report digital copies), AADT queries by road and year, Automatic Road Analyzer (ARAN) (street view logs by year and direction), and ArcGIS data integrated maps application. In addition to the databases, historical Google Earth and Google Maps aerial and street views were used. Intersection related crashes were queried for all nodes within the intersections. Table 2 contains descriptive statistics of the data collected.

The AADT was between 628 and 30,643 vpd over the minor road and between 2,139 and 48,134 vpd over the major road. The posted speed limit ranged from 20 to 45 mph on the minor road and from 25 to 65 mph over the major road. The predominant signal control was protected/permissive. There were, on average, 3.53 exclusive left turn lanes and 1.34 right turn lanes per intersection. On average, there were around two bus stops, 2.46 alcohol sale establishments, and around half of the intersections had one or more educational facility within 1,000 ft of the intersection. The crash data had 8,108 FI, 25,521 PDO, and 33,629 TOT crashes. The descriptive statistics in Table 2 were for the entire database of 601 sites during five years of analysis between 2010 and 2014.

Development of U4SG Calibration Factors

The calibration factor was calculated as the ratio between the total number of observed crashes N_{obs} and the total of predicted crashes N_{pred} using the HSM models across all sites as illustrated in Equation 2 (*1*). The assessment of the quality of the calibration factor consisted of the verification of the magnitude, accuracy, and straight linear relation of observed and predicted crashes.

$$C = \frac{\sum_{all\ sites} N_{obs}}{\sum_{all\ sites} N_{pred}} \tag{2}$$

In Equation 3, the standard deviation of the calibration factor $Sd(C)$ is calculated as the square root of the variance $Var(C)$ which is a function of the observed crashes, predicted crashes, and the overdispersion (k) of the prediction model for site m. In Equations 4 and 5, normal distributed statistics are used with the true calibration factor C_t and the sample calibration factor C to determine the probability P that the sample calibration factor lies within 10% of the true calibration factor in Equation 6—where $\Phi(.)$ is the cumulative density function of the normal distribution. The true calibration factor is derived from the population of sampled sites (*4, 5*). CURE plots as a function of the AADT were also evaluated to verify the calibration factor performance over different ranges.

$$Sd(C) = \sqrt{Var(C)} = \sqrt{\frac{\sum_m \left(N_{obs,m} + k_m N_{obs,m}^2\right)}{\left(\sum_m N_{pred,m}\right)^2}} \tag{3}$$

$$Z_{min} = \frac{0.9 \times C_t - C}{Sd(C)} \tag{4}$$

$$Z_{max} = \frac{1.1 \times C_t - C}{Sd(C)} \tag{5}$$

$$P = \Phi(Z_{max}) - \Phi(Z_{min}) \tag{6}$$

Table 2. Descriptive Statistics of U4SG Intersections

Intersection variable description	Average	St. deviation	Minimum	Maximum
Minor road AADT (vpd)	7,791	5,290	628	30,643
Major road AADT (vpd)	17,781	9,034	2,139	48,134
Posted speed limit minor road (mph)	32.45	5.25	20.00	45.00
posted speed limit major road (mph)	38.58	5.67	25.00	65.00
Signal control				
Legs with permissive signal control	1.05	1.51	0.00	4.00
Legs with protective/permissive signal control	1.87	1.66	0.00	4.00
Legs with protected only signal control	1.07	1.59	0.00	4.00
Right turn on red prohibited	0.11	0.42	0.00	4.00
Exclusive lanes and surrounding area				
Legs with exclusive left turn lanes	3.53	0.95	0.00	4.00
Legs with exclusive right turn lanes	1.34	1.38	0.00	4.00
Bus stops within 1,000 ft.	2.05	2.36	0.00	11.00
Educational facilities present within 1,000 ft	0.43	0.50	0.00	1.00[a]
Alcohol sale establishments within 1,000 ft	2.46	2.18	0.00	8.00
Crash data	Count	Average	St. Deviation	CV
Fatal and injury (FI)	8,108	13.67	10.86	0.79
Property damage only (PDO)	25,521	43.04	37.92	0.88
Total (TOT)	33,629	56.71	46.73	0.82
Number of sites			601	
Period of analysis			2010–2014	

Note: St. deviation = standard deviation.
[a]Some sites had more than one educational facility, the data was coded with 0 when not present and 1 when present.

Since the calibration factor was estimated as the ratio of the observed crashes over the predicted crashes, a linear relationship could be assessed through regression modeling in Equation 7 (*2, 31, 37*).

$$F(C) = N_{obs} = a \times \left(N_{pred}\right)^{b} \tag{7}$$

where:

N_{obs}, observed crashes;
N_{pred}, predicted crashes;
a, model coefficient multiplier (equivalent to calibration factor);
b, model coefficient exponent.

If the model coefficient exponent b is found to be close to the value of 1, there is a straight linear relation between the observed and predicted crashes and the calibration factor should be appropriate for all sites (model coefficient multiplier $a = C$). However, if b were to be different than the value of 1, there is not a straight linear relationship and the calibration would follow a power function based on the predicted crashes by site (calibration function). The Negative Binomial distribution was used for the linear relationship assessment.

Development of U4SG Missouri-Specific Models

Based on the results of the assessment and the large magnitude of the calibration factor, Missouri-specific models were developed. Existing HSM model variables were considered for modeling. Model parameter estimation using maximum likelihood was used. The log-likelihood function based on the Negative Binomial used in model parameter estimation is illustrated in Equation 8 (*37*).

$$\begin{aligned}\ln\left[\mathcal{L}(\beta_0,\beta_1,\ldots,\theta)\right] = \sum_{i=1}^{n}[ln\Gamma(obs_i+\theta) - ln\Gamma(\theta) + \\ \theta\ln(\theta) + obs_i\ln(prrd_i) - \\ (\theta+obs_i)ln(\theta+pred_i)\end{aligned} \tag{8}$$

The letter i denotes units (intersections). The mean incident count for unit i over period of time y_i is u_i. The traits of i define population of units that are assumed to be Gamma distributed with mean $E(u_i)$ and variance $E(u_i)^2/\theta$. The value $1/\theta$ is called the overdispersion parameter which is commonly denoted by the letter k. The parameter estimates of the model coefficients are $\beta_0,\beta_1,\ldots,\theta$. The log-likelihood function that maximizes the estimates are those that maximize the sum of $ln\left[\mathcal{L}(\beta_0,\beta_1,\ldots,\theta)\right]$ (*37*).

The model structure of the HSM for U4SG is a combination of several SPFs (motor vehicle, pedestrian, and bike related crashes) by severity (FI, PDO) and crash type (MV, SV). Pedestrian and bicycle volume counts require a significant data collection effort for modeling. The complexity of the HSM SPFs and applicable CMFs for urban signalized intersections makes its accurate application difficult. For instance, the IHSDM Administration Tool only provides the alternative to calculate the calibration factor for total crashes. Thus, in this paper, the prediction models for severities FI and PDO were developed using data resources available and to facilitate its application. The following predictor variables were considered:

- Minor road AADT (vpd)
- Major road AADT (vpd)
- Minor road posted speed limit (mph)
- Major road posted speed limit (mph)
- Signal control (P, permissive; PP, protected/permissive; PO, protected only)
- No right turn on red
- Number of bus stops within 1,000 ft from the center of the intersection
- Number of alcohol sale establishments within 1,000 ft from the center of the intersection
- Educational facilities present within 1,000 ft from the center of the intersection

Predictor Variables Selection. The methods used for predictor variable selection, order of introduction, and functional form were Exploratory Data Analysis (EDA), Variable Introduction Exploratory Data Analysis (VIEDA), and measures of goodness of fit (*2, 3, 37*). The EDA was used to evaluate the first variable introduced in the model and optimal functional form. In this case, the sum of minor and major road $AADT_{min+maj}$ was introduced first. The data distribution and functional form is illustrated in Figure 1(*a, b*). The optimal functional form was a logistic function shown in Equation 9.

$$F(AADT) =$$

$$F\left(AADT_{min+maj}\right) = \frac{\beta_0}{1+\beta_1 exp\left(\beta_2 \dfrac{AADT_{min+maj}}{10,000}\right)} - \frac{\beta\beta}{1+\beta_1} \quad (9)$$

The VIEDA method was used for visualization of the ratio between the observed crashes and base model predicted crashes (model with some introduced variables) versus values of the variable that was being considered for the model. Bins were generated with specified variable ranges to aggregate the data, estimate the ratios, and look for orderly distributions. Figure 1(*c, d*) illustrates the VIEDA analysis of the variable minor and major road speed limit $S_{min+maj}$. The analysis showed an orderly increasing distribution and the power function shown in Equation 10 was found appropriate. The VIEDA analysis was conducted for all other variables introduced in the model.

$$F(S) = F\left(S_{min+maj}\right) = \beta_3 S_{min+maj}^{\ \beta_4} \quad (10)$$

Measures of Goodness of Fit. The measures of goodness of fit used for model development were the log-likelihood, inverse overdispersion, and CURE plots (*2, 37*). The model parameters that maximized the Negative Binomial likelihood function (Equation 8) are those that maximize the sum of $ln\left[\mathcal{L}(\beta_0,\beta_1,\ldots,\phi)\right]$ resulting in the log-likelihood. An increase in log-likelihood is desired when predictor variables with specified functional forms are introduced in the model. For model improvement, there are many different predictor variables functional forms that were tried (i.e., linear, power, logistic, polynomial, or Hoerl). These functions require different numbers of parameters which may result in parameter inflation. Thus, the Akaike Information Criterion (AIC) was used to determine if the improvement in fit was genuine. The overdispersion parameter indicates the variability of the model in comparison with a Poisson distribution with the same mean. The reliability of the resulting models is likely to be higher with a larger value of the inverse overdispersion coefficient ($\phi = 1/k$). CURE plots track model performance throughout the range of predictor variables. A satisfactory CURE plot is one that follows a symmetric random walk around the horizontal axis and within the confidence intervals ($\mp 2\sigma$). In contrast, sudden vertical changes represent outliers, and long increasing or decreasing runs represent regions of consistent under- or overprediction (*37*). Throughout the process of adding more variables, trying different functional forms, or changing the order in which the variables were introduced, CURE plots were continuously evaluated.

Results

This section provides the results of the development and assessment of the calibration factor and jurisdiction-specific model for U4SG intersections in Missouri.

U4SG Calibration Factor

The calibration included all sampled sites (601 intersections) with 3 years of data during 2012–2014. The resulting calibration factor was 3.98 with a standard deviation of 0.13 (99.68% confidence level). The magnitude of the calibration factor was much larger than the value of 1 showing a significant difference between the observed data in Missouri and the HSM models. The straight linear relationship assumption between observed and predicted crashes was also assessed. The model coefficients were $a = 0.82$ and $b = 1.68$ (Equation 11). Since the coefficient b was different than the value of 1, there was not a straight linear relationship between observed and predicted crashes.

$$F(C) = N_{obs} = 0.82 \times \left(N_{pred}\right)^{1.68} \quad (11)$$

To assess the calibration results, Figure 2(*a*) presents observed versus predicted (not calibrated) crashes and graphical representations of calibration factors and calibration function. The data points were found to be significantly displaced from the HSM model estimates (black dashed line, C

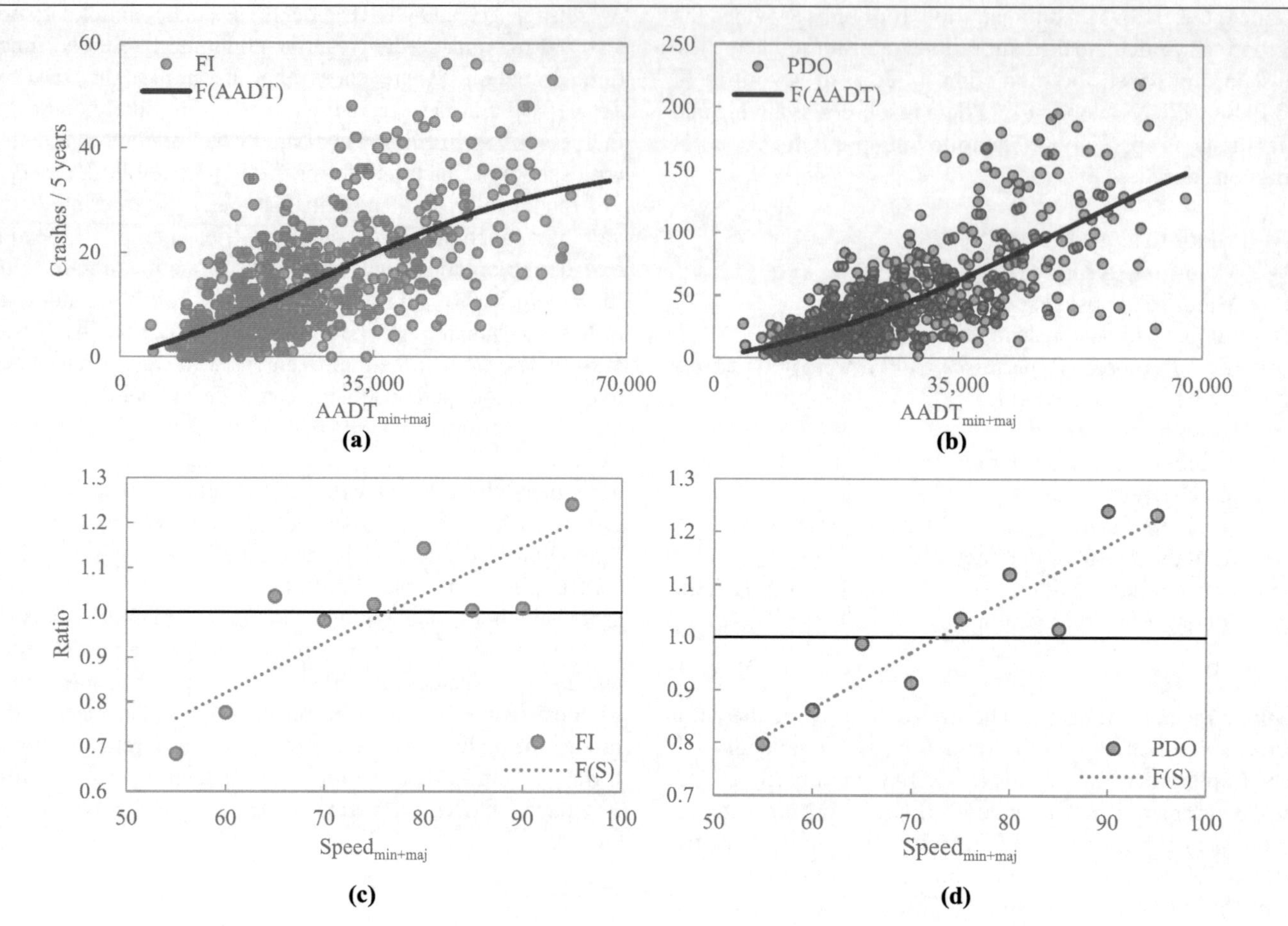

Figure 1. (a, b) EDA analysis with AADT and (c, d) VIEDA analysis with posted speed limit by severity.

= 1.00). The calibration factor and function fit the observed versus predicted crash data points, but this does not mean that the magnitude and accuracy of the estimates were properly represented. It is important to assess the quality of the calibrated estimates with predictor variables other than the fitted values. Figure 2(*b*) presents CURE plots of the calibration factor and function in relation to the intersection AADT. Both calibration factor and function did poorly along all ranges of AADT consistently under- or overpredicting and going beyond the confidence limits ($\mp 2\sigma$). Thus, the calibration of HSM models (U4SG) to Missouri conditions was not appropriate because of the clear difference in the observed data and the model predictions. Thus, the development of jurisdiction-specific models was supported.

U4SG Missouri Jurisdiction-Specific Model

The estimated Missouri-specific prediction model for U4SG intersections is provided in Equation 12. Table 3 provides the model coefficients for FI and PDO crashes as each predictor variable was introduced in the model. The final model coefficients were **highlighted in red** in the last column for Model 10.

$$N_{U4SG,y} = \left[\frac{\beta_0}{1+\beta_1 exp\left(\beta_2 \frac{AADT_{min+maj}}{10{,}000}\right)} - \frac{\beta_0}{1+\beta_1} \right] \times \beta_3 S_{min+maj}{}^{\beta_4} \times (\beta_5 P + \beta_6 PP + \beta_7 PO) \times (\beta_8 LL + 1) \times (\beta_9 RL + 1) \times \beta_{10}{}^{I_{PPRa}} \times \left[\beta_{11} exp(\beta_{12} BST)\right]^{I_{BST}} \times (\beta_{13} ALC + 1) \times \beta_{14}{}^{I_{SCH}} \times \varphi \quad (12)$$

where:

$N_{U4SG,y}$, number crashes per year for U4SG and severities y (FI and PDO);

$AADT_{min+maj}$, sum of minor and major road AADT;

$S_{min+maj}$, sum of minor and major road posted speed limit;

P, permsive signal control;

PP, protected/permissive signal control;

PO, protected only signal control;

LL, number of approach legs with exclusive left turn lanes;

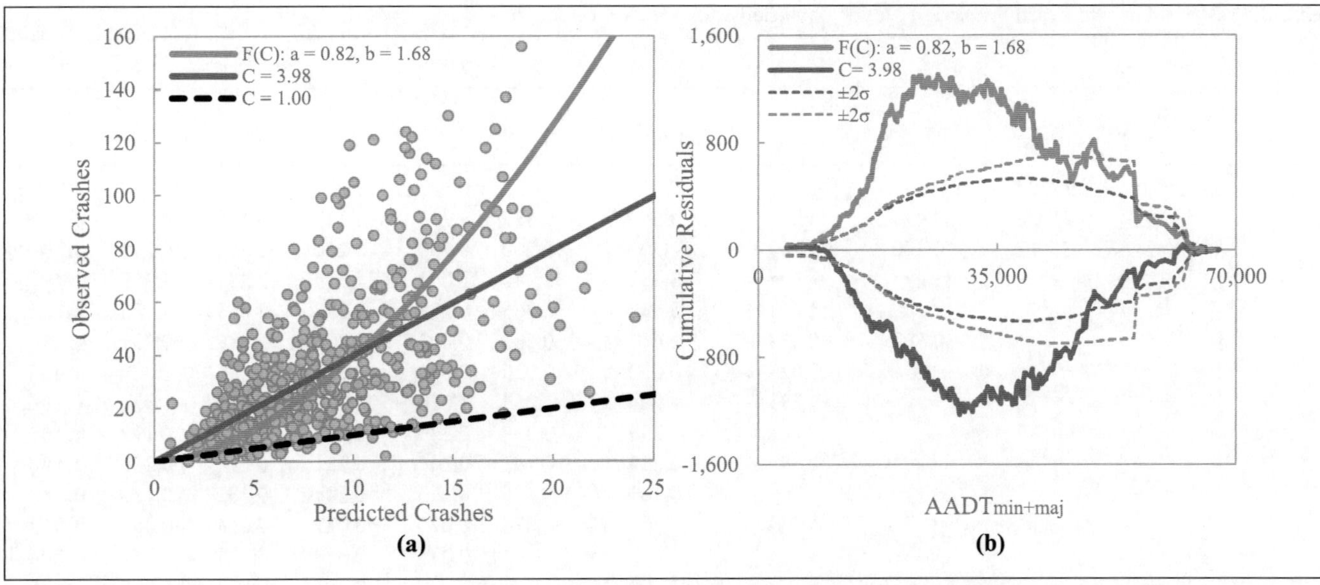

Figure 2. (a) Observed and predicted crashes plot and (b) CURE plots as function of AADT.

RL , number of approach legs with exclusive right turn lanes;

BST , number of bus stops within 1,000 ft from the intersection;

ALC , number of alcohol sale establishments within 1,000 ft from the intersection;

I_{PROH} , 1 when one or more legs right turn is prohibited, 0 otherwise;

I_{BST} , 1 when bus stops present, 0 otherwise;

I_{SCH} , 1 when educational facilities present within 1,000 ft from the intersection, 0 otherwise;

β_{0-14} , model coefficients;

φ , crashes per year adjustment factor.

For application, the models at different stages of the development can be used based on data availability. For instance, if the AADT and posted speed limits were the only data available, Model 3 coefficients could be used. The sequence of introduction of variables showed a consistent magnitude in the model coefficients. All predictor variables contributed to the model in terms of measures of goodness of fit. The AADT had the most significant contribution to the models with an improvement of 166.7 (FI) and 241.8 (PDO) in log-likelihood. Similarly, the improvement in inverse overdispersion was of 1.723 (FI) and 1.897 (PDO). The predictor variables BST and ALC had significant contributions to the models with 26.3 (FI) and 23.7 (PDO) in log-likelihood. The posted speed limit also made a significant contribution to the model in both log-likelihood and inverse overdispersion. The fully loaded model CURE plots as a function of AADT and posted speed limit are presented in Figure 3. The walk of cumulative residuals showed optimal oscillations within the confidence intervals and around the value of zero at all ranges of AADT and posted speed limit.

The reader should be careful not to overinterpret individual coefficients in Table 3, as the goal of the modeling process was to produce overall estimates of crash frequency. It is not recommended to make causal inferences with safety cross-sectional models (*24*). It is difficult to determine whether the effect of a variable is strongly dependent on the specific modeling conditions or whether the regression failed to capture cause and effect. Instead, unbiased observational studies are recommended for this purpose.

Conclusions

The HSM provides a prediction methodology composed of crash prediction models that estimate the expected number of crashes per year, by facility type, severity, and crash type. There are two approaches to apply the HSM crash prediction methodology to local conditions: calibration of existing models in the HSM, or development of jurisdiction-specific crash prediction models. There are some instances in which the calibration of HSM models may not be recommended. The calibration factor for U4SG intersections for Missouri conditions was 3.98 with a standard deviation of 0.13. The assessment of the calibration factor showed a disproportional difference between the observed data in Missouri and the HSM model for U4SG intersections. Thus, the calibration of HSM models for U4SG intersections to Missouri conditions was deemed inappropriate and the development of jurisdiction-specific models was supported.

Missouri-specific crash prediction models for severities FI and PDO were developed. The predictor variables considered were intersection AADT, posted speed limit, signal

Table 3. Missouri-Specific Crash Prediction Model Coefficients for U4SG Intersections

Severity	Model	1	2	3	4	5	6	7	8	9	10
	Variable introduced	None	$AADT_{min+maj}$	$S_{min+maj}$	$P, PP, and\ PO$	LL	RL	I_{PROH}	BST, I_{BST}	ALC	I_{SCH}
FI	Log-lik.	14599.0	14765.8	14772.5	14778.1	14782.1	14783.1	14784.4	14810.7	14814.4	**14816.5**
	𝒷	1.702	3.425	3.546	3.640	3.692	3.712	3.727	4.189	4.262	**4.298**
	φ	2.735	0.200	0.200	0.200	0.200	0.200	0.200	0.200	0.200	**0.200**
	β_0		46.272	47.624	33.452	33.447	35.042	37.304	17.231	14.955	**12.799**
	β_1		4.688	4.113	5.492	6.365	6.238	6.287	3.863	3.585	**3.721**
	β_2		−0.562	−0.612	−0.612	−0.666	−0.692	−0.706	−0.738	−0.751	**−0.773**
	β_3			0.029	0.020	0.020	0.021	0.023	0.011	0.009	**0.008**
	β_4			0.793	1.004	1.064	1.020	0.983	1.274	1.327	**1.396**
	β_5				0.259	0.239	0.249	0.252	0.196	0.193	**0.185**
	β_6				0.232	0.242	0.246	0.251	0.218	0.208	**0.198**
	β_7				0.189	0.196	0.195	0.201	0.182	0.171	**0.163**
	β_8					−0.074	−0.078	−0.079	−0.066	−0.068	**−0.067**
	β_9						0.035	0.038	0.045	0.041	**0.044**
	β_{10}							0.859	0.892	0.898	**0.885**
	β_{11}								1.125	1.130	**1.125**
	β_{12}								0.061	0.056	**0.055**
	β_{13}									0.036	**0.037**
	β_{14}										**1.105**
	Increase Log-lik.		166.7	6.7	5.7	4.0	1.1	1.2	26.3	3.7	2.1
	Increase 𝒷		1.723	0.121	0.094	0.053	0.019	0.015	0.463	0.072	0.036
PDO	Log-lik.	77643.2	77885.0	77893.8	77906.1	77908.8	77914.2	77915.0	77923.0	77946.7	**77948.2**
	𝒷	1.539	3.436	3.549	3.716	3.748	3.824	3.834	3.942	4.317	**4.337**
	φ	8.607	0.200	0.200	0.200	0.200	0.200	0.200	0.200	0.200	**0.200**
	β_0		251.098	257.299	409.842	397.664	408.713	417.924	357.855	275.092	**242.543**
	β_1		8.434	6.289	3.391	3.760	3.218	3.237	2.867	1.865	**1.834**
	β_2		−0.437	−0.444	−0.305	−0.346	−0.397	−0.418	−0.424	−0.397	**−0.401**
	β_3			0.029	0.046	0.045	0.046	0.047	0.040	0.031	**0.027**
	β_4			0.783	0.556	0.601	0.523	0.501	0.562	0.661	**0.717**
	β_5				0.238	0.222	0.235	0.236	0.219	0.219	**0.211**
	β_6				0.306	0.313	0.317	0.318	0.307	0.286	**0.275**
	β_7				0.367	0.373	0.360	0.366	0.355	0.323	**0.311**
	β_8					−0.058	−0.066	−0.067	−0.059	−0.064	**−0.062**
	β_9						0.075	0.077	0.074	0.063	**0.066**
	β_{10}							0.894	0.892	0.903	**0.894**
	β_{11}								0.792	0.797	**0.792**
	β_{12}								0.064	0.056	**0.056**
	β_{13}									0.099	**0.099**
	β_{14}										**1.079**
	Increase Log-lik.		241.8	8.8	12.3	2.7	5.4	0.8	8.0	23.7	1.5
	Increase 𝒷		1.897	0.113	0.167	0.032	0.076	0.010	0.108	0.376	0.019

Note: 𝒷 is the inverse overdispersion ($1/k$). The fully loaded model coefficients are for Model 10 (coefficients in bold).

control type, exclusive left turn lanes, exclusive right turn lanes, right turn on red prohibited, and facilities of interest within 1,000 ft from the intersection (bus stops, schools, and alcohol sale establishments). The posted speed limit predictor variable provided a significant contribution to the models. The HSM models do not explicitly include speed limit as a predictor variable. The log-likelihood, inverse overdispersion, and CURE plots showed satisfactory measures of model accuracy. The models developed in this paper should be used in crash analysis and coefficients magnitudes are not intended for independent variable interpretation.

There is wide range of results with the implementation of HSM models from different jurisdictions. This paper serves as a case study in which the calibration of the HSM for U4SG was inappropriate. A thorough assessment of the calibration is recommended to determine the quality of the

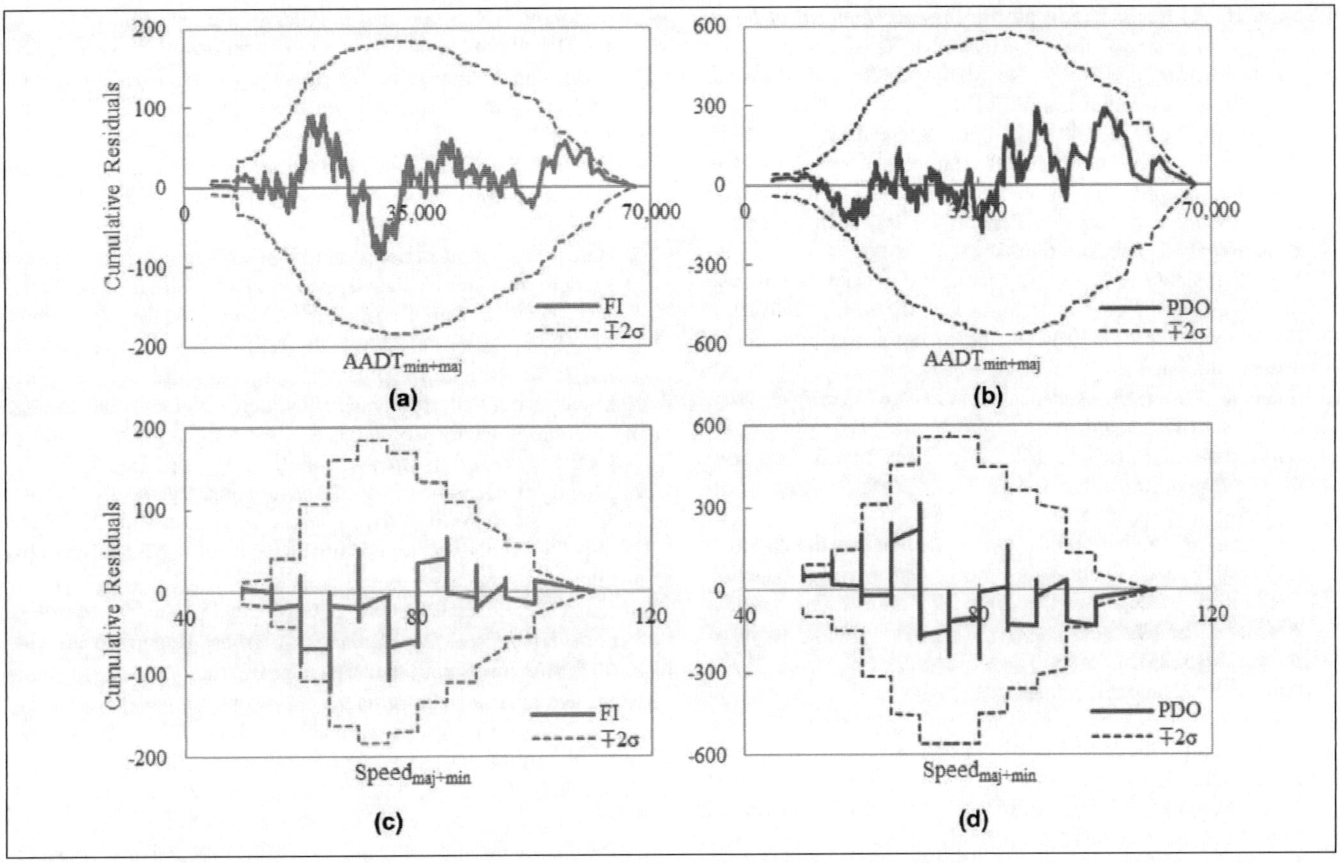

Figure 3. CURE plots as a function of (a, b) AADT and (c, d) posted speed limit.

estimates. Jurisdictions should develop their own prediction models by carefully assembling consistent databases in which sites share comparable and identifiable features and optimizing the functional forms of predictor variables in the model.

References

1. American Association of State Highway Transportation Officials. *Highway Safety Manual (HSM) with Supplement*, 1st ed. Washington, DC: AASHTO, 2014.
2. Srinivasan, R., D. Carter, and K. Bauer. Safety Performance Function Decision Guide: SPF Calibration vs SPF Development. *Report No. FHWA-SA-14-004*. FHWA, U.S. Department of Transportation, 2013.
3. Srinivasan, R., and K. Bauer. *Safety Performance Function Development Guide: Developing Jurisdiction-Specific SPFs*. Report No. FHWA-SA-14-005. FHWA, U.S. Department of Transportation, 2013.
4. Bahar, G. *HR 20-7(332): User's Guide to Develop Highway Safety Manual Safety Performance Function Calibration Factors*. National Cooperative Highway Research Program, Transportation Research Board, Washington, D.C., 2014.
5. Shirazi, M., D. Lord, and S. Geedipally. Sample-size Guidelines for Recalibrating Crash Prediction Models: Recommendations for the Highway Safety Manual. *Accident Analysis and Prevention*, No. 93, 2016, pp. 160–168.
6. Banihashemi, M. Sensitivity Analysis of Data Set Sizes for Highway Safety Manual Calibration Factors. *Transportation Research Record: Journal of the Transportation Research Board*, 2012. 2279: 75–81.
7. Alluri, P., D. Saha, and A. Gan. Minimum Sample Sizes for Estimating Reliable Highway Safety Manual (HSM) Calibration Factors. *Journal of Transportation Safety and Security*, Vol. 8, No. 1, 2016, pp. 56–74.
8. Trieu, V., S. Park, and J. McFadden. Use of Monte Carlo Simulation for a Sensitivity Analysis of Highway Safety Manual Calibration Factors. *Transportation Research Record: Journal of the Transportation Research Board*, 2014. 2435: 1–10.
9. Dixon, K., C. Monsere, R. Avelar, J. S. Barnett, and P. Escobar. *Improved Safety Performance Functions for Signalized Intersections*. Report No. FHWA-OR-RD-16-03. Oregon Department of Transportation, 2015.
10. Minnesota Department of Transportation. *Pavement Management: Video-log*. MnDOT, http://www.dot.state.mn.us/materials/pvmt-mgmt.html. Accessed July 22, 2017.
11. Missouri Department of Transportation. *MoDOT's Automatic Roadway Analyzer (ARAN)*. MoDOT. http://www.modot.org/expresslane/ARAN.htm. Accessed July 22, 2017.
12. Srinivasan, S., P. Haas, N. S. Dhakar, R. Hormel, D. Torbic, and D. Harwood. *Development and Calibration of Highway Safety Manual Equations for Florida Conditions*. Report No. TRC-FDOT-82013-2011. Florida Department of Transportation, 2011.

13. Shin, H., Y. Lee, and S. Dadvar. *The Development of Local Calibration Factors for Implementing the Highway Safety Manual in Maryland*. Report No. MD-14-SP209B4J. Maryland State Highway Administration, 2014.
14. Savolainen, P. T., T. Gates, D. Lord, S. Geedipally, E. Rista, T. Barrette, B. J. Russo, and R. Hamzeie. *Michigan Urban Trunkline Intersections Safety Performance Functions (SPFs) Development and Support*. Report No. RC-1628. Michigan Department of Transportation, 2015.
15. Sun, C., P. Edara, H. Brown, B. Claros, and K. Nam. *Calibration of the Highway Safety Manual for Missouri, Addendum*. Report No. cmr14-007 (addendum). Missouri Department of Transportation, 2014.
16. Shankar, V., and S. Madanat. *Methods for Identifying High Collision Concentrations for Identifying Potential Safety Improvements: Development of Safety Performance Functions for California*. Report No. CA 15-2317. California Department of Transportation, 2015.
17. Troyer, D., K. Bradbury, and C. Juliano. Strength of the Variable: Calculating and Evaluating Safety Performance Function Calibration Factors for the State of Ohio. *Transportation Research Record: Journal of the Transportation Research Board*, 2015. 2515: 86–93.
18. American National Standards Institute. *Manual on Classification of Motor Vehicle Traffic Accidents*, 7th ed. ANSI D16.1-2007. ANSI, 2007.
19. Wang, X., M. Abdel-Aty, A. Nevarez, and J. Santos. Investigation of Safety Influence Area for Four-legged Signalized Intersections: Nationwide Survey and Empirical Inquiry. *Transportation Research Record: Journal of the Transportation Research Board*, 2008. 2083: 86–95.
20. Green, E. R., C. Blackden, and M. A. Fields. *Spatial Database for Intersections*. Report No. KTC-15-11/SPR14-480-1F. Federal Highway Administration, 2015.
21. Harwood, D., K. Bauer, K. Richard, D. Gilmore, J. Graham, I. Potts, D. Torbic, and E. Hauer. *NCHRP Project 17-26: Methodology to Predict the Safety Performance of Urban and Suburban Arterials*. Transportation Research Board of the National Academies, Washington, D.C., 2007.
22. Persaud, B., D. Lord, and J. Palmisano. Calibration and Transferability of Accident Prediction Models for Urban Intersections. *Transportation Research Record: Journal of the Transportation Research Board*, 2002. 1784: 57–64.
23. Sacchi, E., B. Persaud, and M. Bassani. Assessing International Transferability of Highway Safety Manual Crash Prediction Algorithm and its Components. *Transportation Research Record: Journal of the Transportation Research Board*, 2012. 2279: 90–98.
24. Hauer, E. Cause, Effect and Regression in Road Safety: A Case Study. *Accident Analysis and Prevention*, Vol. 42, No. 4, 2010, pp. 1128–1135.
25. Kononov, J., C. Lyon, and B. Allery. Relation of Flow, Speed, and Density of Urban Freeways to Functional Form of a Safety Performance Function. *Transportation Research Record: Journal of the Transportation Research Board*, 2011. 2236: 11–19.
26. Hauer, E. Statistical Safety Modeling. *Transportation Research Record: Journal of the Transportation Research Board*, 2004. 1897: 81–97.
27. Hauer, E., F. Council, and Y. Mohammedshah. Safety Models for Urban Four-Lane Undivided Road Segments. *Transportation Research Record: Journal of the Transportation Research Board*, 2004. 1897: 96–105.
28. Lu, J., K. Haleem, P. Alluri, A. Gan, and K. Liu. Developing Local Safety Performance Functions Versus Calculating Calibration Factors for SafetyAnalyst Applications: A Florida Case Study. *Safety Science*, Vol. 65, 2014, pp. 93–105.
29. Minnesota Department of Transportation. *Phase I-II of the Minnesota Highway Safety Manual Calibration*. MnDOT, 2014. http://www.dot.state.mn.us/trafficeng/safety/docs/mnhsmcalibration.pdf. Accessed July 22, 2017.
30. Sun, C., P. Edara, H. Brown, B. Claros, and K. Nam. *Calibration of the Highway Safety Manual for Missouri*. Report No. cmr14-007. Missouri Department of Transportation, 2013.
31. Srinivasan, R., and D. Carter. *Development of Safety Performance Functions for North Carolina*. Report No. FHWA/NC/2010-09. North Carolina Department of Transportation, 2011.
32. Smith, S., D. Carter, and R. Srinivasan. *Updated and Regional Calibration Factors for Highway Safety Manual Crash Prediction Models*. Report No. FHWA/NC/2016-09. North Carolina Department of Transportation, 2017.
33. Xie, F., K. Gladhill, K. Dixon, and C. Monsere. Calibration of Highway Safety Manual Predictive Models for Oregon State Highways. *Transportation Research Record: Journal of the Transportation Research Board*, 2011. 2241: 19–28.
34. Persaud and Lyon Inc. *Safety Performance Functions for Intersections*. Report No. CDOT-2009-10. Colorado Department of Transportation, 2009.
35. Tegge, R. A., J. Jo, and Y. Ouyang. *Development and Application of Safety Performance Functions for Illinois*. Report No. FHWA-ICT-10-066. Illinois Department of Transportation, 2010.
36. Donnell, E. T., V. V. Gayah, and L. Li. *Regionalized Safety Performance Functions*. Report No. FHWA-PA-2016-001-PSU WO 017. Pennsylvania Department of Transportation, 2016.
37. Hauer, E. *The Art of Regression Modeling in Road Safety*. Springer, New York, 2016.

The Standing Committee on Highway Safety Performance (ANB25) peer-reviewed this paper (18-01754).

TRR
JOURNAL OF THE TRANSPORTATION RESEARCH BOARD

Article

Transportation Research Record
2018, Vol. 2672(30) 43–52
© National Academy of Sciences:
Transportation Research Board 2018
Article reuse guidelines:
sagepub.com/journals-permissions
DOI: 10.1177/0361198118758011
journals.sagepub.com/home/trr
SAGE

Before–After Evaluation of the Realignment of Horizontal Curves on Rural Two-Lane Roads

Raghavan Srinivasan[1], Daniel Carter[1], Craig Lyon[2], and Matthew Albee[3]

Abstract

This study determined the crash modification factors (CMFs) associated with horizontal curve realignment using the before–after empirical Bayes method and compared the results from published CMFs from cross-sectional studies. This evaluation used data from rural, two-lane roads in California, North Carolina, and Ohio. The evaluation revealed a 68% reduction in total crashes, a 74% reduction in injury and fatal crashes, a 78% reduction in run-off-road and fixed object crashes, a 42% reduction in crashes during dark conditions, and an 80% reduction in wet-road crashes, all of which were statistically significant at the 95% confidence level. The results pertain to a range of site characteristics, the most important of which is the range of before and after degree of curve. The average degrees of curve in the before and after periods were 18.1 (with a minimum of 3.2 and a maximum of 52.1) and 6.9 (with a minimum of 0.0 and a maximum of 16.3), respectively. The average central angle of the curves was approximately 42° (with a minimum of 1 and a maximum of 117). The CMFs from this before–after evaluation are lower compared to CMFs estimated from two previous cross-sectional studies. There is a need for further research with a larger sample of sites to assess the reliability of the CMFs obtained from this before–after evaluation.

Horizontal curves have consistently been an area of concern for road safety practitioners, particularly for single-vehicle crashes. A 2010 report from the Bureau of Transportation Statistics showed that one-third of fatal single-vehicle crashes occur at curves (*1*). Most crashes at curves involve a vehicle leaving the roadway and striking a fixed object or overturning (*2, 3*).

The American Association of State Highway and Transportation Officials *Guide for Reducing Collisions on Horizontal Curves* lists "increasing the radius of a horizontal curve" as one strategy for reducing the likelihood of a vehicle leaving its lane (*3*). From 2013 to 2015, 23 states reported using Highway Safety Improvement Program funds to implement horizontal curve realignment via 83 projects (*4*).

Typically, researchers have conducted studies on the effect of horizontal curve radius using cross-sectional regression analysis, where they compare the safety performance of curves of different radii. This methodology allows for a larger potential sample of curves because there is not a need to know a conversion/installation date such as is needed in a before–after study. It allows the use of an entire state's database of curved road sections, as was the case with Washington State data in two recent studies (*5, 6*). However, there are many possible issues with such cross-sectional regression methods, including the "difficulty to properly control for unknown, or known but unmeasured factors, omitted variable bias, and correlation of variables" (*7*). On the other hand, Elvik (*8*) indicates that well-designed before–after studies are less subject to confounding and more likely to produce reliable crash modification factors (CMFs) because in a before–after study, "we are dealing with the same roadway unit located in a particular place used by probably the same users in the before and after period." One difficulty with before–after studies is identifying where and when the safety improvement was implemented. With the support of state departments of transportation (DOTs) from California, Ohio, and North Carolina, the

[1]UNC Highway Safety Research Center, Chapel Hill, NC
[2]Persaud and Lyon, Inc., Ottawa, ON, Canada
[3]VHB, Inc., Raleigh, NC

Corresponding Author:
Address correspondence to Raghavan Srinivasan: srini@hsrc.unc.edu

authors obtained detailed information on curve realignment projects on rural two-lane roads, allowing for a before–after evaluation. This evaluation involved the analysis of the following crash types:

- Total;
- Injury and fatal;
- Run-off-road and fixed object;
- Crashes during dark; and
- Wet-road crashes.

Literature Review

Several studies have produced CMFs and crash modification functions (CMFunctions) for curve radius on rural, two-lane roads. The *Highway Safety Manual, First Edition* (HSM) chapter for rural, two-lane roads includes a CMF for horizontal curvature (p. 10–27) (*9*). The HSM based this CMF on regression analysis done by Zegeer et al. (*10*). The base condition of the HSM predictive model for rural, two-lane roads is a tangent, and the CMF for a curve is calculated as

$$CMF = \frac{(1.55L) + \left(\frac{80.2}{R}\right) - (0.012S)}{1.55L}$$

where

L = length of horizontal curve (mi), including spiral transitions;

R = radius of curve (Ft); and

S = 1 if spiral transition present, 0 if not, 0.5 if present only on one end of curve.

A study by Bauer and Harwood focused on the interaction of horizontal curvature and vertical curvature and grade for rural, two-lane highways (*5*). The authors conducted a cross-sectional analysis of roadway segments from Washington State. They used 3,457 mi of road and six years of crash data to produce a series of CMFunctions that quantify the safety performance of various combinations of vertical and horizontal alignment.

Their study classified horizontal alignment of roadway segments as follows:

- Tangent
- Horizontal curve

Their study classified vertical alignment of roadway segments as follows:

- Level
- Straight grade (constant grade of 1% or more)
- Type 1 or 2 crest vertical curve
- Type 1 or 2 sag vertical curve

One of the CMFunctions produced by Bauer and Harwood (*5*) is

$$CMF = \exp\left[0.044G + 0.19\ln\left(2\times\frac{5730}{R}\right) + 4.52\left(\frac{1}{R}\right)\left(\frac{1}{L_c}\right)\right]$$

where G is absolute value of % grade, R is the curve radius (Ft), and L_c is the horizontal curve length (mi). This function calculates a CMF for fatal and injury crashes for a horizontal curve that occurs on a straight grade (i.e., no vertical curvature). The study produced other CMFunctions for other conditions (e.g., horizontal curve on a type 1 vertical crest).

Saleem and Persaud (*6*) built on the earlier studies on two-lane highways to explore the development of a CMFunction where they explored different designs to flatten an existing horizontal curve. They focused on horizontal curves only on level grade (less than 3%). They used 440 curves from Washington State with radii between 100 and 11,000 ft to conduct a cross-sectional analysis and develop CMFunctions. Their functions show that CMFs decreased (fewer crashes occurred) for scenarios in which a larger radius is entertained. They furthermore found that the level of annual average daily traffic (AADT) on the subject segment did not influence the safety effect.

Other studies have produced CMFs related to horizontal curve radius but focused on other types of crash or roadway conditions:

- Schneider et al. estimated CMFs for horizontal curvature on two-lane highways, but focused on truck-involved crashes (*11*).
- Banihashemi, Fitzpatrick et al., and Graham et al. produced CMFs involving horizontal curve radius but focused on rural, multilane highways (*12–14*).
- Pratt et al. and Choi et al. produced CMFs for horizontal alignment but focused on freeways (*15, 16*).

Summary of Literature Review

Previous studies have generally shown that larger curve radius is associated with fewer crashes. However, the vast majority of the published studies that have examined the safety effect of horizontal curvature were based on cross-sectional studies that may not always provide reliable CMFs. This illustrates the need for before–after studies.

Overview of Methodology

This evaluation used the empirical Bayes (EB) methodology for observational before–after studies. This method is currently well established in the field of highway safety and is well documented in multiple publications (e.g., *17*, *18*). This method accounts for the possible bias due to regression to the mean (RTM) and properly accounts for changes in traffic volume during the study period.

The following steps are needed to conduct an EB before–after evaluation:

1. Identify a reference group of sites without the treatment, but similar to the treatment sites in terms of the major factors that affect crash risk including traffic volume and other site characteristics.
2. Estimate Safety Performance Functions (SPFs) using data from the reference sites relating crashes to independent variables, such as traffic volume and other site characteristics.
3. In estimating SPFs, calibrate annual SPF multipliers to account for the temporal effects (e.g., variation in weather, demography, and crash reporting) on safety. The annual SPF multiplier is the ratio of the observed crashes to the predicted crashes from the SPF. In using the annual SPF multipliers from the SPFs to account for temporal effects, it is assumed that the trends in the crash counts are similar in the treatment and reference groups.
4. Use the SPFs, annual SPF multipliers, and data on traffic volumes for each year in the before period for each treatment site to estimate the number of crashes that would be predicted for the before period in each site.
5. Calculate the EB estimate of the expected crashes in the before period at each treatment site as the weighted sum of the actual crashes in the before period and predicted crashes from step 4.
6. For each treatment site, estimate the product of the EB estimate of the expected crashes in the before period and the SPF predictions for the after period divided by these predictions for the before period. This is the EB expected number of crashes that would have occurred had there been no treatment. The variance of this expected number of crashes is also estimated in this step. The expected number of crashes without the treatment along with the variance of this parameter and the number of reported crashes after the treatment is used to calculate the CMF and the standard deviation of the CMF.

When planning a before–after safety evaluation study, it is vital to ensure that enough data are included to statistically detect the expected change in safety. Even though at the planning stage the expected change in safety is not known, it is still possible to make a rough determination of how many sites the study will require, based on the best available information about the expected change in safety. For a detailed explanation of sample size considerations, as well as estimation methods, see chapter 9 of Hauer (*17*). In this study, the methods from Chapter 9 of Hauer (*17*) were used to determine the number of sites that are needed to statistically detect the expected change in safety.

Data

Treatment and Reference Sites

The initial list of potential treatment sites was assembled based on information obtained from the states' safety engineers in addition to the State Safety Improvements Project Database developed as part of the Highway Safety Information System (HSIS). The focus of this evaluation was on projects involving an increase in the horizontal radius. Among the states that provided data on curve realignments, the focus was on sites in California, North Carolina, and Ohio, because these states represented some of the largest sample sizes, and all three of these states contribute data to HSIS. The use of HSIS data was an advantage not only because it facilitates the process of obtaining the crash and roadway data, but also because it enabled the team to obtain annual estimates for the traffic volume.

Reference sites were identified by visually inspecting the roadway network at the area of each treatment curve and selecting one or two other curves nearby on the same route. As far as possible, the curves with small radii were selected to match the "before" condition of the treatment sites. Aerial imagery was inspected to ensure that no significant changes had taken place at these reference curves. Similarly, aerial imagery and reports from the states were reviewed to ensure that no other significant changes had taken place at the treatment sites apart from the realignment. The timeline of aerial imagery was also used to verify (and occasionally correct) the installation year for the treatment sites.

Road Characteristics and Traffic Volume Data

HSIS was used to obtain data on roadway characteristics and traffic volume. Although the states provided location data for the treated curves, it was often general information, such as the milepost range for an entire roadway project, or simply a description (e.g., "SR 1100 from 0.4 mile to 0.2 mile west of US 13 Bypass, west of Windsor"). To obtain more precise information of the location of the sites, the treatment and reference sites were first identified in Google Earth, then overlaid on a geographic information system (GIS)-based road

Table 1. Site Characteristics Summary Statistics

Site characteristic	Range of values in data for treatment sites	Range of values in data for reference sites
Range of curve radius (Ft) before realignment in treatment sites and in reference sites	110 to 1,780	75 to 2,900
Average curve radius (Ft) before realignment	430	600
Range of curve radius (Ft) after realignment in treatment sites	350 to 4,500 (and essentially straight tangent for four sites with radii over 10,000)	N/A
Average curve radius (Ft) after realignment (including radius over 10,000 feet)	2,700	N/A
Average curve radius (Ft) after realignment (excluding radius over 10,000 feet)	1,000	N/A
Range of shoulder width (Ft)	1 to 8	1 to 9
Average shoulder width (Ft)	3.9	3.7
Range of AADT (veh/day)	440 to 12,000	400 to 12,000
Average AADT (veh/day)	3,480	2,600

network from each state. The GIS road file was used to identify the start and end mileposts of each curve and to verify or obtain other pieces of data needed to identify the road section, such as route number and county. In some cases, state engineers were contacted to obtain data that could not be obtained from the GIS road file and other sources.

The following information was collected from the GIS road files and aerial imagery (Google Earth):

- Curve radius before and after—A GIS-based tool was used to measure the radius of the curve before and after realignment. Typically, the GIS linework represented either the before or after alignment (depending on how recently the realignment occurred and how up-to-date the GIS file was), and the team used aerial imagery to obtain a measurement of the other period.
- Distance to previous and next curve—The distance from the start of the study curve to the endpoint of the next curve, and the distance from the end of the study curve to the start point of the next curve, were noted. These data were used to distinguish isolated curves from grouped curves.
- Intersection in curve—Some curves had an intersection of a public road in the curve section or very near the endpoints of the curve. This information was noted as well.

Table 1 provides a summary of characteristics for both treatment and reference sites.

Crash Data

Crash data were obtained from HSIS. Crashes were assigned to each segment according to the start and end mileposts of the curve (i.e., the point of curvature and point of tangency). In relying on mileposts, the authors recognize the potential for data error, specifically crash locating, when working with sections of small length. In addition, crash positions in rural areas can vary in accuracy because of fewer landmarks and intersections.

For treatment sites, the realignment of the curve from a sharp curve to a flatter curve resulted in a longer effective length of the curve in the after period. Mileposts from the after period for crash data were used to determine the boundaries of the sites. This implies that crashes from the before period for the treatment site would include some amount of tangent section on either end of the curve. The authors considered an alternative approach which would limit crashes in the before period to only those crashes occurring on the curve. However, that would mean a comparison from before to after periods of two different lengths—a shorter length in the before period and a longer length in the after period. The authors felt this alternative approach would have an inherent bias because of the length difference. Table 2 provides the summary statistics regarding the crash data.

Safety Performance Functions

SPFs were estimated using generalized linear modeling, assuming a negative binomial error distribution. In specifying a negative binomial error structure, the team iteratively estimated the dispersion parameter, k, from the model and the data. Due to the relatively small sample sizes from the individual states, data for the three states were combined to estimate the SPF with a separate intercept term estimated for each state. Only the SPF for total crashes was estimated. The SPFs for the other crash types could not be reliably estimated, probably because of their

Table 2. Crash Data Summary Statistics

Variables	Treatment sites before	Treatment sites after	Reference sites
Number of sites	39	39	56
Years of data	220	130	610
Mile-years	33.2	18.8	55.8
Total crashes	198	29	214
Injury and fatal crashes	109	10	86
Run-off-road plus fixed object crashes	132	11	104
Dark crashes	61	13	53
Wet crashes	88	7	78

limited sample sizes. To predict the number of crashes for the other crash types, the authors applied the SPF for total crashes with a multiplier equal to the number of crashes of the type of interest divided by the number of total crashes at the reference sites. The following multipliers were obtained for each crash type:

- Injury and fatal: 0.40
- Run-off-road and fixed-object: 0.49
- Head-on: 0.02
- Sideswipe: 0.08
- Dark: 0.25
- Wet: 0.36

As mentioned earlier, to assess the entire impact on safety of curve realignment to a larger radius curve, it is necessary to consider a study area that extends beyond the limits of the smaller radii curve. When changing from a smaller to a larger radius, the reconstruction removes tangents on either side of the curve. The study area therefore needs to include the entire roadway travelled to get from the beginning to ending of the largest curve radius under consideration.

Because the reference group consisted of only curved segments, the SPF predictions had to be adjusted to consider the expected crash frequencies on the tangent segments associated with the smaller radius curve in the before period that were subsequently removed. For this adjustment, the SPF calibrated from the reference group was applied to the tangent segments as if the tangent segments removed had the same curvature as the curved segment to which they were adjacent. Then, this estimate was adjusted by an estimated ratio of expected crashes on a tangent to expected crashes on a curve. The adjustment made use of the model developed by Zegeer et al., which is the basis of the CMF for horizontal curvature on two-lane roads in the HSM (*9,10*). The model predicts the number of crashes per year as

$$A = 1.55L + 0.014D - 0.012S) \times C \times 0.978^{w-30}$$

where

A = expected crash frequency,
L = length of the curve (mi),
D = degree of curvature,
S = 1 if spirals exist and 0 otherwise,
C = volume of vehicles (in millions), and
w = roadway width (Ft).

It was unknown if spiral transitions existed at the treated curves, so it was assumed that spiral transitions are not present. Recognizing that the traffic volume and road width remain unchanged and the degree of curvature for a tangent is 0, the adjustment was estimated as

$$\frac{A_{\text{tangent}}}{A_{\text{curve}}} = \frac{1.552L}{1.552L + 0.014D}$$

where

L = total tangent length, and
D = degree of curve of the curved segment adjacent to the tangents.

The estimate of crashes for the before period conditions, which include both the curved and tangent segments removed, is then

$$\text{Crashes per year} = (\text{SPF applied to curve segment}) + \left(\frac{A_{\text{tangent}}}{A_{\text{curve}}}\right)(\text{SPF applied to tangent segments removed})$$

Below is the form of the SPF developed using the reference group sites, the parameters of which are presented in Table 3.

$$\textit{Crashes per year} = \exp[\text{a} + \text{b} + \text{c} \times \ln(\text{AADT}) + \text{d} \times \ln(\text{Length}) + \text{e} \times \text{RAD}]$$

where

AADT = annual average daily traffic volume,
Length = length of horizontal curve (mi),
RAD = radius of horizontal curve (mi),
a, b, c, d, e = parameters estimated in the SPF calibration process (a is the intercept, b represents a set of coefficients for each state),
ln is the natural logarithm function, and
exp is the exponential function.

Table 3. SPF Parameter Estimates with Standard Errors

Parameter	Value	Standard error
a	−0.8951	1.6329
b (California)	−1.2198	0.3352
b (North Carolina)	−0.7103	0.3881
b (Ohio: reference level)	0.0000	na
c	0.4424	0.1509
d	0.9659	0.2943
e	−3.7331	1.6331
k *(overdispersion parameter)*	0.4527	0.1555

Results

Aggregate Analysis

Table 4 provides the estimates of expected crashes in the after period without treatment, the observed crashes in the after period, and the estimated CMF and its standard error for all crash types considered. The CMFs reported in Table 4 are based on all 39 locations that the states realigned. CMFs were also estimated after removing the three locations that the states realigned to straight segments, and the resulting CMFs were almost identical to the ones reported in Table 4. The results indicate large reductions that are statistically significant at the 95% confidence level for all crash types analyzed.

The results pertain to a range of site characteristics, the most important of which is the range of before and after curve radii. The average degrees of curve in the before and after periods were 18.1 and 6.9 respectively (the degree of curve is mentioned here instead of radius because three of the curves were realigned to straight segments, which have undefined radius values but have a degree of 0.0). The average central angle of the curves was approximately 42°, and the average AADT at these sites was about 3,500. The average length of the segments the states realigned was about 0.15 miles.

Comparison with results from cross-sectional studies. One of the intents of this evaluation was to compare the results of the before–after evaluations with published results from cross-sectional studies. Two approaches were used to do this comparison. The first approach involved entering the average radii in the before and after periods along with the average segment length of the study sites to estimate the CMFs based on the equations given in Zeeger et al. (*10*) (for total crashes) and Bauer and Harwood (*5*) (for injury and fatal crashes). Based on the equation given in Zegeer et al., and assuming no spiral transition curves, the ratio of the CMF for a curve with a degree 6.9 to a curve with degree 18.1 is 0.672 for total crashes. For the same conditions, the CMFunction in Bauer and Harwood estimate a CMF of 0.784 for injury and fatal crashes.

The first approach is simplistic in that it does not specifically account for the fact that the relationship between the CMF and curve radius and segment length is nonlinear in both Zegeer et al. (*10*) and Bauer and Harwood (*5*). To address this limitation, a second approach was identified. The second approach involved the following steps:

1. For each treated site, calculate the CMF based on Zegeer et al. (*10*) and Bauer and Harwood (*5*).
2. Multiply the EB expected crashes in the after period for each site (this is the expected crashes had the treatment not been implemented) with the CMFs from Zegeer et al., (*10*) and Bauer and Harwood (*5*). This product gives the expected crashes with the treatment for each site based on the CMFs from these two studies.
3. Calculate the sum of the expected crashes with the treatment for the all sites based on the two studies.
4. The ratio of the expected crashes with the treatment to the EB expected crashes without the treatment provides an approximate CMF associated with the equations from the two studies.

Based on the second approach, the CMF based on the equations in Zegeer et al. (*10*) was 0.699 (for total crashes), and the CMF based on the equations in Bauer and Harwood (*5*) was 0.736 (for injury and fatal crashes). These CMFs are higher than the CMFs of 0.315 and

Table 4. Aggregate CMF Results

Metric	Total	Injury and fatal	Run-off-road plus fixed-object	Dark	Wet-road
EB estimate of crashes expected in after period without strategy	91.43	38.11	50.29	21.90	33.88
Count of crashes observed in after period	29	10	11	13	7
Estimate of CMF	**0.315**	**0.259**	**0.216**	**0.584**	**0.204**
Standard error of estimate of CMF	0.064	0.086	0.068	0.176	0.079

0.259 that were estimated using the before and after EB evaluation for total and injury and fatal crashes, respectively. There is some evidence that cross-sectional studies may indicate smaller crash reductions (i.e., higher CMFs) compared to before–after studies. Gross et al. (*7*) report that "CMFs from cross-sectional designs tend to indicate smaller expected crash reductions than those derived from before–after studies". For example, a recent evaluation of the pedestrian hybrid beacon found that the CMF based on an EB before–after evaluation was 0.24, but the CMF based on a cross-sectional regression model was about 0.53 (*19*).

Disaggregate Analysis

The expected CMF value for a specific site could vary significantly based on its characteristics. The focus of the disaggregate analysis was on total crashes due to the small number of after period crashes from other crash types. Since the CMF could be a function of the site characteristics (e.g., change in the radius or degree of curvature due to realignment), an attempt was made to further analyze the data to develop a CMFunction for the purposes of estimating a site-specific CMF.

The traditional approach for estimating CMFunctions includes the use of the CMF value as the dependent variable and site/treatment characteristics as independent variables:

$$\mathrm{CMF} = f(\textit{site characteristics})$$

where, *f* represents a generic function. This CMFunction could then be estimated as a regression equation. Elvik (*20*) recommended that the variance of the CMF needs to be considered in this estimation. The inverse of the variance is typically introduced as a weight in a weighted regression model. In other words, for an observation (or site) whose CMF is $\mathrm{CMF_i}$ with a variance of $\mathrm{Var(CMF_i)}$, the weight will be $1/\mathrm{Var(CMF_i)}$. For linear regression, this would be appropriate.

Some recent studies have recommended the use of a different model form such as a log-normal model that would ensure the predicted CMF from a CMFunction would always be greater than zero. In the case of the log-normal model, it has been shown that the appropriate weight for a weighted log-normal regression model would instead be $[\mathrm{CMF_i}/\mathrm{Var(CMF_i)}]$ (*21*).

For either the normal regression or log-normal regression models with weights, reliable estimates of CMFs and their variances are necessary. To have reliable estimates of these parameters, researchers often combine sites with similar characteristics, especially if there are sites where the reported crashes in the after period are 0. However, this aggregation can lead to loss of useful information. In this study, a different approach was used to overcome the disadvantage of losing information due to aggregation. This approach involves rewriting the CMFunction as

$$\mathrm{CMF_i} = \frac{\delta_i}{\pi_i} = f(\textit{site characteristics})$$

where

$\pi_i =$ expected number of crashes that would have occurred in the after period without the strategy at site *i*, and

$\delta_i =$ number of reported crashes in the after period at site *i*.

The CMFunction above can be rewritten as

$$\delta_i = \pi_i \times f(\textit{site characteristics})$$

Written in this form, it is possible to estimate this equation as a count data model (such as Poisson or negative binomial model) with δ as the dependent variable and π as the offset. This approach is similar to estimating an SPF with the goal of predicting the number of crashes per mile, where crash frequency is included as the dependent variable and section length is included as the offset. One limitation with this approach is that the offset is an estimated value from the EB procedure with a variance, and hence, not a fixed measure value. There has been some limited research in traffic safety on the implications of such variance in the independent variables, but further research is necessary (e.g., *22*).

The independent variables for the CMFunction included the following:

- AADT
- Difference/ratio between the after and before radius/degree of curvature
- Central angle of the curve

The functional form for the CMFunction that was estimated using negative binomial regression is

$$\mathrm{CMF} = \exp(a_0 + a_1X_1 + a_2X_2 + \ldots a_nX_n)$$

where a_0 represents the intercept, and a_1 through a_n represent the coefficients for independent variables X_1 through X_n.

In three locations, states aligned the curved sections to be a straight segment, indicating a radius of infinity in the after period. However, expressing the curvature as a degree of the curve (5,730/radius in feet) instead of radius would allow this study to include these sites the states realigned to a straight segment, since the degree of curvature for a straight segment is 0. To address this issue, the project team explored two options:

- Option 1: Include all 39 sites

Table 5. Option 2 CMFunction for Total Crashes

Variable	Estimate (Standard Error)
intercept	−3.4775 (3.7253)
ln(AADT)	0.5299 (0.4219)
central angle	−0.0342 (0.0162)
radius after - radius before (mi)	−1.6628 (1.5026)
overdispersion parameter (*k*)	0.3951 (0.5239)

- Option 2: Exclude the 3 sections that the states realigned to a straight segment (i.e., include data from 36 sites)

Multiple CMFunctions were considered, including the ratio of degree of curvature instead of the difference in degree of curvature from the before to the after periods (in option 1), and the ratio of the radius (and/or curvature) instead of the difference in radius (and/or curvature) from the before and after periods (in option 2). The best CMFunction was from option 2, which is presented in Table 5.

The estimated parameters indicate that the CMF increases as AADT increases but decreases as the difference between the radii of the two curves increases and as the central angle increases. Many of the parameters have a high standard error including the overdispersion parameter. If the true value of the overdispersion parameter were 0 then a poisson model would be appropriate. The same model was calibrated with a poisson error distribution but showed little change in the parameter estimates.

Due to the high standard errors, this CMFuction cannot be considered as robust. However, the general indications of the CMFunction may be useful in a contemplated curve realignment. In particular, safety benefits may be greater for curves with a larger central angle and where the difference in radius between the before and after period conditions is larger. This is consistent with previous cross-sectional studies (*5,10*) that indicate that the safety benefits due to realignment would be greater with a larger increase in radius from the before to the after periods.

Economic Analysis

The following steps were undertaken for the economic analysis:

1. Using the number of mile-years in the after period, the change in crashes per mile-year were determined. The expected benefit due to the realignment was estimated as 1.495 injury and fatal crashes per mile per year and 1.826 property damage only (PDO) crashes per mile per year.
   a. Note: The change in PDO crashes was estimated using the EB expected crashes (for total and injury and fatal crashes) in the after period and the actual crashes in the after period (for total and injury and fatal crashes).
2. For the benefit calculations, the most recent Federal Highway Administration (FHWA) mean comprehensive crash costs disaggregated by crash severity and location type was used as a base (*23*). Based on 2001 crash costs and the unit cost (in 2001 dollars), the cost for injury and fatal crashes and PDO crashes in rural areas was \$206,015 and \$7,800 respectively. This was updated to 2015 dollars by applying the ratio of the U.S. Department of Transportation (U.S. DOT) 2015 value of a statistical life of \$9.4 million to the 2001 value of \$3.8 million (U.S. DOT 2015) (*24*). Applying this factor of 2.48 to the unit costs resulted in an aggregate 2015 unit cost of \$510,446 for injury and fatal crashes and \$19,326 for PDO crashes. The expected annual benefit due to the fewer crashes after realignment was \$510,446 × 1.495 + \$19,326 × 1.826, which totals approximately \$798,507. Based on guidance in the U.S. DOT document (*24*), sensitivity analysis was conducted by estimating a low value for the benefit as 0.55 × \$798,507 and a high value for the benefit as 1.38 × \$798,507. The low and high values of the benefit estimates were used to estimate the low and high values for the benefit cost ratio.
3. The annualized cost of the treatment was estimated as

$$\textit{Annual Cost} = \frac{C \times i}{1 - (1 + i)^{-N}}$$

where

C = treatment cost. Based on information from North Carolina, the average realignment cost per mile was \$3,121,599;

i = discount rate (as a decimal) and assumed to be 0.07; and

N = expected service life (years) of 30 years.

The annualized cost per year for realignment was \$251,558.

4. The project team calculated the benefit cost ratio of the annual crash savings to the annualized treatment cost. The benefit cost ratio was 3.17:1 with a low of 1.75:1 to a high of 4.38:1.

Conclusions and Limitations

This evaluation used the EB before–after method to evaluate the safety aspects of horizontal curve realignment on rural, two-lane roads, based on data from North Carolina, Ohio, and California. Not surprisingly, the sample size in this evaluation was limited because realignments are expensive. However, previous studies used cross-sectional regression methods that may not always provide reliable CMFs.

The evaluation revealed a 68% reduction in total crashes, a 74% reduction in injury and fatal crashes, a 78% reduction in run-off-road and fixed object crashes, a 42% reduction in crashes during dark conditions, and an 80% reduction in wet-road crashes, all of which were statistically significant at the 95% confidence level. The average degrees of curve in the before and after periods were 18.1 (with a minimum of 3.2 and a maximum of 52.1) and 6.9 (with a minimum of 0.0 and a maximum of 16.3), respectively. The average central angle of the curves was approximately 42° (with a minimum of 1 and a maximum of 117). Estimated CMFunctions showed that safety benefits may be greater for curves with a larger central angle and where the difference in radius between the before and after period conditions is larger. The economic analysis revealed a benefit cost ratio of 3.17:1 with a range of 1.75:1 to 4.38:1.

This study compared the total crash CMFs with the results from previous cross-sectional studies. The CMFs from this before–after evaluation are lower compared to CMFs estimated from two previous cross-sectional studies.

Limitations

There are several recognized limitations to these findings:

- The sample size is limited because realignments are expensive. Further studies with a larger set of sites may be useful.
- Grade and vertical curvature were not available for this evaluation. As shown in the work by Bauer and Harwood (*5*), the presence of a grade or vertical curve can affect the CMF of the change in horizontal curve radius.
- Information on the presence or absence of a spiral transition for each curve was not available, though the total curve length and crash data would have included any spiral portion of road. The work by Zegeer et al. (*10*) and presented in the HSM (*9*) includes a factor for spiral presence.
- This evaluation did not include information about chevrons or other types of delineation at the horizontal curves that were realigned. However, other previous cross-sectional studies (*5*) also did not account for delineation of the curves.
- The authors recognize the potential for data error, specifically crash locating, when working with sections of small length. In addition, crash positions in rural areas especially can vary in accuracy. Further discussion of this issue is provided in Appendix B of an upcoming FHWA report (*25*).

Acknowledgments

This paper is based on research sponsored by Federal Highway Administration, Office of Safety Research and Development under Contract DTFH61-13-D-0001. The FHWA program and task manager for this project was Roya Amjadi. The authors would like to thank research assistants Kari Signor and Jen Scholz from UNC HSRC for their help in compiling the data. The authors thank the Highway Safety Information System (HSIS) for providing the roadway and crash data for California, Ohio, and North Carolina.
The team gratefully acknowledges that this research would not have been possible without the assistance provided by the following agencies:

- California Department of Transportation
- North Carolina Department of Transportation
- Ohio Department of Transportation

The authors would like to especially thank the North Carolina Department of Transportation staff for their detailed review of the crash data from North Carolina and for educating the authors about the importance of crash location errors in short sections.

References

1. Bureau of Transportation Statistics–2010. http://www.rita.dot.gov/bts/sites/rita.dot.gov.bts/files/publications/bts_technical_report/september_2010/index.html. Accessed May, 2016.
2. Srinivasan, R., J. Baek, D. Carter, B. Persaud, C. Lyon, K. Eccles, F. Gross, and N. Lefler. *Safety Evaluation of Improved Curve Delineation.* Report No. FHWA-HRT-09-045. Federal Highway Administration, Washington, DC, 2009.
3. Torbic, D., D. Harwood, D. Gilmore, R. Pfefer, T. Neuman, K. Slack, and K. Hardy, *Guidance for Implementation of the AASHTO Strategic Highway Safety Plan*, Volume 7: A Guide for Reducing Collisions on Horizontal Curves, National Cooperative Highway Research Program, Report 500. Transportation Research Board, Washington, DC, 2004.
4. Highway Safety Improvement Program. *Database of state projects.* FHWA, unpublished, 2016.
5. Bauer, K. M., and D. W. Harwood. *Safety Effects of Horizontal Curve and Grade Combinations on Rural Two-Lane Highways.* Report FHWA-HRT-13-077. Federal Highway Administration, Washington, DC, 2014.
6. Saleem, T., and Persaud, B. Another Look at the Safety Effects of Horizontal Curvature on Rural Two-Lane

Highways. Presented at 95th Annual Meeting of the Transportation Research Board, Washington, D.C., 2016.
7. Gross, F., B. Persaud, and C. Lyon. *A Guide to Developing Quality Crash Modification Factors*. Report FWHA-SA-10-032. Federal Highway Administration, Washington, DC, 2010.
8. Elvik, R. Assessing Causality in Multivariate Accident Models. *Accident Analysis and Prevention*, Vol. 43, 2011, pp. 253–264.
9. Highway Safety Manual. 1st ed. American Association of State Highway and Transportation Officials, Washington, DC, 2010.
10. Zegeer, C., J. Stewart, F. Council, D. Reinfurt, and E. Hamilton. Safety Effects of Geometric Improvements on Horizontal Curves. *Transportation Research Record: Journal of the Transportation Research Board*, 1992. 1356: 11–19.
11. Schneider, W. H., K. Zimmerman, D. Van Boxel, and S. Vavilikolanu. A Bayesian Analysis of the Effect of Horizontal Curvature on Truck Crashes Using Training and Validation Data Sets. *Transportation Research Record: The Journal of the Transportation Research Board*, 2009. 2096: 41–46.
12. Banihashemi, M. Is Horizontal Curvature a Significant Factor of Safety in Rural Multilane Highways? *Transportation Research Record: The Journal of the Transportation Research Board*, 2015. 2515: 50–56.
13. Fitzpatrick, K., D. Lord, and B. Park. Horizontal Curve Accident Modification Factor with Consideration of Driveway Density on Rural, Four-Lane Highways in Texas. *Journal of Transportation Engineering*, Vol. 136, No. 9, 2010.
14. Graham, J., D. Harwood, K. Richard, M. O'Laughlin, E. Donnell, and S. Brennan. *NCHRP Report 794: Median Cross-Section Design for Rural Divided Highways*. Transportation Research Board of the National Academies, Washington, D.C., 2014.
15. Pratt, M., S. Geedipally, A. Pike, P. Carlson, A. Celoza, and D. Lord. *Evaluating the Need for Surface Treatments to Reduce Crash Frequency on Horizontal Curves*. Report No. FHWA/TX-14/0-6714-1. Texas Department of Transportation. Austin, Texas, 2014.
16. Choi, Y., S. Kho, C. Lee, and D. Kim. Development of Crash Modification Factors of Alignment Elements and Safety Countermeasures for Korean Freeways. Presented at 94th Annual Meeting of the Transportation Research Board, Washington, D.C., 2015.
17. Hauer, E. *Observational Before-After Studies in Road Safety*, Pergamon Press, 1997.
18. Persaud, B., and C. Lyon. Empirical Bayes Before-After Safety Studies: Lessons Learned from Two Decades of Experience and Future Directions, *Accident Analysis and Prevention*, Vol. 39, No. 3, 2007, pp. 546–555.
19. Zegeer, C., C. Lyon, R. Srinivasan, B. Persaud, B. Lan, S. Smith, D. Carter, N. J. Thirsk, J. Zegeer, E. Ferguson, R. Van Houten, and C. Sundstrom, Development of Crash Modification Factors for Uncontrolled Pedestrian Crossing Treatments. *Transportation Research Record: The Journal of the Transportation Research Board*, 2017. 2636: 1–8.
20. Elvik, R. Methodological Guidelines for Developing Accident Modification Functions, *Accident Analysis and Prevention*, Vol. 80, 2015, pp. 26–36.
21. Bonneson, J. *Local adjustment of CMFs based on crash distribution*, Working Paper 3, NCHRP Project 17–63 (Guidance for the Development and Application of Crash Modification Factors). (In press).
22. Wright, D. E., and V. Barnett. *Fitting Predictive Accident Models in GLIM with Uncertainty in Flow Estimates*. Contractor Report 286, Transport and Road Research Laboratory, 1991.
23. Council, F., E. Zaloshnja, T. Miller, and B. Persaud. *Crash Cost Estimates by Maximum Police-Reported Injury Severity Within Selected Crash Geometries*, FHWA-HRT05-051, Federal Highway Administration, Washington, DC, 2005.
24. USDOT. *Guidance on Treatment of the Value of a Statistical Life (VSL) in U.S. Department of Transportation Analyses—2015 Adjustment*, USDOT. Available at: https://cms.dot.gov/sites/dot.gov/files/docs/VSL2015_0.pdf. Accessed May 10, 2016.
25. Srinivasan, R., D. Carter, C. Lyon, and M. Albee. *Safety Evaluation of Horizontal Curve Realignment on Rural Two-Lane Roads*, FHWA-HRT-17-066, Federal Highway Administration, Washington, DC, 2017.

The Standing Committee on Highway Safety Performance (ANB25) peer-reviewed this paper (18-01803).

TRR
JOURNAL OF THE TRANSPORTATION RESEARCH BOARD

Article

Transportation Research Record
2018, Vol. 2672(30) 53–63
© National Academy of Sciences:
Transportation Research Board 2018
Article reuse guidelines:
sagepub.com/journals-permissions
DOI: 10.1177/0361198118796352
journals.sagepub.com/home/trr
SAGE

Modified Mixed Generalized Ordered Response Model to Handle Misclassification in Injury Severity Data

Lacramioara Balan[1] and Rajesh Paleti[2]

Abstract

Traditional crash databases that record police-reported injury severity data are prone to misclassification errors. Ignoring these errors in discrete ordered response models used for analyzing injury severity can lead to biased and inconsistent parameter estimates. In this study, a mixed generalized ordered response (MGOR) model that quantifies misclassification rates in the injury severity variable and adjusts the bias in parameter estimates associated with misclassification was developed. The proposed model does this by considering the observed injury severity outcome as a realization from a discrete random variable that depends on true latent injury severity that is unobservable to the analyst. The model was used to analyze misclassification rates in police-reported injury severity in the 2014 General Estimates System (GES) data. The model found that only 68.23% and 62.75% of possible and non-incapacitating injuries were correctly recorded in the GES data. Moreover, comparative analysis with the MGOR model that ignores misclassification not only has lower data fit but also considerable bias in both the parameter and elasticity estimates. The model developed in this study can be used to analyze misclassification errors in ordinal response variables in other empirical contexts.

The World Health Organization (WHO) considers road safety as a major public health problem given that nearly 1.25 million people lose their lives and another 50 million people sustain nonfatal injuries every year globally (*1*). Additionally, traffic crashes remain the leading cause of death within the 15 to 29 years age group. In the United States alone, 35,000 people lost their lives and 2.44 million people were injured in 2015 (*2*). To address this problem, safety engineers undertake data analysis to identify policy measures to enhance roadway safety. In the United States, within each state, traffic accidents are investigated by police officers who complete a standard form, usually soon after a crash has occurred, called a police accident report (PAR). The report contains information about driver characteristics, vehicle attributes, traffic conditions, environment conditions, and crash characteristics (*3*). Typically, all accidents that are above a certain severity level and a threshold property damage dollar value are recorded by police (*4*). These PARs constitute the primary data component used for safety analysis. For instance, the General Estimates System (GES) data of the National Automotive Sampling System (NASS) is a nationally representative sample of police-reported crashes across the nation. Although the PARs are mostly reliable, there are several factors that determine the quality of data recorded. Consistency of coverage and interpretation, missing data, response errors, entry procedure, and level of detail are some of the factors identified in the literature (*5*).

A key crash attribute used for safety analysis is the injury severity of the crash. The crash injury severity level is recorded as the severity level of the most severely injured person in the crash. In some cases, the PARs record the severity level of injuries sustained by all the people involved in the crash. Typically, injury severity is recorded based on crash-assigned or hospital-assigned ordinal scoring systems (*6*). The crash-assigned injury severity reported in PARs is typically recorded on a five-point ordinal scale—fatal, incapacitating, non-incapacitating, possible, and no injury. However, the definitions of these ordinal categories in the KABCO scale vary across state jurisdictions. The hospital-assigned injury severity, on the other

[1]Civil & Environmental Engineering Department, Old Dominion University, Norfolk, VA
[2]Department of Civil and Environmental Engineering, The Pennsylvania State University, University Park, PA

Corresponding Author:
Address correspondence to Rajesh Paleti: rpaleti@engr.psu.edu

hand, is an anatomically based Abbreviated Injury Scale (AIS) that rates an injury on a six-point scale (minor, moderate, serious, severe, critical, and maximum) based on the threat to life and is correlated with mortality, morbidity, and hospital stay duration (*4*).

Whereas some studies found significant differences between the KABCO and AIS scoring systems (*4*), others found that these two measures are reasonably consistent (*6*). For instance, a comparison of police-reported crashes and hospital records in New Zealand found that police recorded only two-thirds of fatal crashes. Furthermore, the reporting rates were found to vary by demographic, crash, seasonal, and geographic factors (*7*). Along similar lines, pedestrian nonfatal crashes were found to be under-reported in police records (*8*). A similar comparison between police records and trauma registry in France found misclassification in all injury severity categories (*9*). Similar under-reporting and misclassification of injury severity of crashes leading to medical care was found in other recent studies (*10–13*). But to the contrary, police records have also been found to considerably overestimate injury severity and the degree of overestimation was found to vary by the injury severity score (ISS) and the victim's age and position inside the vehicle (*3*). Overall, past research suggests that hospital-based AIS recordings are more precise compared with the KABCO scale police recordings. However, there is no consensus on the exact level of discordance, the nature of discordance, and the factors that lead to discordance between these two measures of injury severity (*8, 9, 14, 15*).

From a modeling standpoint, these data errors are referred as "mismeasurement" errors and more specifically as "misclassification" errors in the case of discrete responses. In the case of linear models (e.g., linear regression), ignoring the errors in the dependent variable will lead to unbiased but less precise parameter estimates (*16*). However, in the case of limited dependent variable models (e.g., discrete choice models), the standard maximum likelihood estimation that ignores misclassification will lead to biased and inconsistent parameter estimates. In fact, even a 2% misclassification rate was found to result in 15% to 25% bias in the parameter estimates in the binary choice scenario (*17*). So, even small errors in the injury severity data can have huge implications for the accuracy of predictions and policy sensitivity analysis. Whatever type of injury severity scoring system is used, there may be errors in crash databases (*18, 19*). For instance, multiple injury severity patterns can have the same score in AIS depending on the weights associated with different body parts. Furthermore, the error rates in the police injury severity recordings that form the basis for most safety research are expected to be higher compared with hospital-based severity scores. However, though there have been attempts to measure the errors in crash databases by comparing them with alternate data sources (such as hospital and ambulance records), none of these studies attempted to account for these errors in statistical models used for analyzing injury severity.

In this context, the objective of the current research is to develop a statistical model to analyze police-reported injury severity while accounting for potential misclassification errors by building on the existing literature in econometrics. Specifically, this study adopted the modified maximum likelihood inference method that treats the observed injury severity outcomes as realizations from discrete random variables that depend on true latent injury severity outcomes that are not observed by the analyst. The misclassification rates constitute the probability mass associated with different possible values of the observed outcome for a given true latent severity level (*17*). An important point note there is that the methodology adopted in this paper can be used to estimate errors relative to truth (which is unknown) in the same injury severity scoring system (but not relative to other data sources). For instance, although there are guidelines on recording the severity outcomes of crashes, in some cases, there are borderline events that warrant some subjective judgments by the police officers at the crash site. Moreover, another source of misclassification in police-recorded data is that some bodily injuries can be latent and might be wrongly coded as less severe crashes. This study measures the extent of these errors in police-recorded severity data relative to the unknown true KABCO rating that crash should have received. The empirical analysis was undertaken using the 2014 GES database maintained by the NHTSA. However, it is important to note that the proposed model (and this research) does not offer insights into the possible reasons for the misclassification errors but rather quantifies and adjusts for the bias caused by these errors. Moreover, the study does not model under-reporting in the total number of crashes (*20–22*) but rather misclassification across multiple injury severity categories. This study adopted the mixed generalized ordered probit (MGORP) framework for modeling injury severity outcomes conditional on crash occurrence. A brief description of the model is given in the next section.

Methodological Framework

The ordered response (OR) framework assumes a single latent propensity function that is mapped into one of J ordered outcomes by $J-1$ threshold parameters that are strictly ordered. The latent propensity function is specified as the sum of a linear-in-parameters deterministic component (which is a function of observed attributes) and a random component (that represents all the unobserved factors that influence the ordered outcome). The specification of the OR model is completed by assuming a continuous probability density function for the random component. The two most commonly used assumptions for the density function of the unobserved component are the standard normal distribution (leading to the ordered response probit [ORP] model) and the standard logistic distribution (leading to the ordered response logit [ORL] model) (*23, 24*). Earlier applications of the OR models assumed constant threshold parameters that do not vary across observations. However, for the same reasons that the latent propensity function varies across observations, the threshold parameters can vary systematically across observations. This idea led to the formulation of the generalized ordered response (GOR) framework that parameterized thresholds as a function of observation-specific attributes (*25*). The next important extension of the OR framework is capturing the "unobserved heterogeneity" of parameters in the propensity and threshold components; in other words, the effects of different observed attributes can vary across observations because of the moderating influence of unobserved factors not considered in the model (*26*). Researchers developed the mixed GOR (MGOR) model that assumes the parameters in the propensity and threshold components to be stochastic realizations from multivariate probability density functions to address this problem (*25, 27*). This study adopted the mixed generalized ordered probit (MGORP) framework for modeling injury severity outcomes conditional on crash occurrence.

Let $q(1,2,\ldots.Q)$ be the index for the crash and $j(1,2,\ldots J)$ be the index for the injury severity outcome. In the current context, $J = 4$ with the four ordered alternatives being no injury ($j = 1$), possible injury ($j = 2$), non-incapacitating injury ($j = 3$), and incapacitating or fatal injury ($j = 4$). In the OR framework, the latent propensity y_q^* is related to the $K \times 1$ vector of observed attributes $\boldsymbol{x}_q$ (including constant) as

$$y_q^* = \boldsymbol{\beta}_q' \boldsymbol{x}_q + \varepsilon_q \quad (1)$$

where $\boldsymbol{\beta}_q$ is the vector of parameters corresponding to the observed attributes $\boldsymbol{x}_q$ and ε_q is the stochastic component of propensity assumed to be a realization from the standard normal distribution, such that $\varepsilon_q \sim N(0,1)$. The subscript q to the parameter vector indicates unobserved heterogeneity across observations. The $\boldsymbol{\beta}_q$ vector is assumed to be a realization from a multivariate normal distribution with mean $\boldsymbol{b}$ vector and $K \times K$ covariance matrix $\boldsymbol{\Sigma}$, such that $\boldsymbol{\beta}_q \sim N(\boldsymbol{b}, \boldsymbol{\Sigma})$. Equation 1 can now be re-written as

$$y_q^* = \boldsymbol{b}'\boldsymbol{x}_q + \tilde{\boldsymbol{\beta}}_q' \boldsymbol{x}_q + \varepsilon_q = \boldsymbol{b}'\boldsymbol{x}_q + \eta_q \quad (2)$$

where $\tilde{\boldsymbol{\beta}}_q \sim N(0_K, \boldsymbol{\Sigma})$ and 0_K is a $K \times 1$ vector of zeros. The variance of effective error term η_q is equal to $\boldsymbol{x}_q' \boldsymbol{\Sigma} \boldsymbol{x}_q + 1$.

The latent propensity y_q^* is mapped into ordinal outcomes by threshold parameters ψ_q^k as

$$y_q = j \ if \ \psi_q^{j-1} < y_q^* < \psi_q^j \quad (3)$$

The strict monotonicity of thresholds is ensured by using the following parameterization:

$$\psi_q^k = \psi_q^{k-1} + \exp\left(\boldsymbol{\gamma}_{k,q}' z_{k,q}\right), \psi_q^0 = -\infty, \psi_q^J = \infty, \text{ and } \psi_q^1 = \exp(\alpha_1) \quad (4)$$

where α_1 is a scalar parameter that corresponds to the first threshold $z_{k,q}$ is an $L_k \times 1$ vector of observed attributes affecting the kth threshold and $\boldsymbol{\gamma}_{k,q}$ is the corresponding vector of coefficients which is assumed to be a stochastic realization from a multivariate normal distribution with mean c_k and $L_k \times L_k$ covariance matrix $\boldsymbol{\Omega}_k$, such that $\boldsymbol{\gamma}_{k,q}' \sim N(\boldsymbol{c}_k, \boldsymbol{\Omega}_k)$. Please note that the observed attributes $\boldsymbol{x}_q$ can include a constant because in Equation 4 all the thresholds are constrained to be positive using the exponential parameterization.

Let $\boldsymbol{\theta}_q = \left(\boldsymbol{\gamma}_{1,q}', \boldsymbol{\gamma}_{2,q}', \ldots \boldsymbol{\gamma}_{J-1,q}'\right)'$ and $\boldsymbol{c} = (\boldsymbol{c}_1', \boldsymbol{c}_2', \ldots.\boldsymbol{c}_{J-1}')'$ denote $\left(\sum_{k=1}^{J-1} L_k\right) \times 1$ vectors of vertically stacked parameters $\boldsymbol{\gamma}_{k,q}$ and $\boldsymbol{c}_k$. The probability of ordinal outcome j conditional on random parameter vectors $\boldsymbol{\gamma}_{k,q}$ in thresholds can be obtained by

$$\begin{aligned} P(y_q = j|\boldsymbol{\theta}_q) &= P\left(\psi_q^{j-1} < y_q^* < \psi_q^j\right) \\ &= P\left(\psi_q^{j-1} < \boldsymbol{b}'\boldsymbol{x}_q + \eta_q < \psi_q^j\right) \\ &= P\left(\psi_q^{j-1} - \boldsymbol{b}'\boldsymbol{x}_q < \eta_q < \psi_q^j - \boldsymbol{b}'\boldsymbol{x}_q\right) \\ &= \Phi\left(\frac{\psi_q^j - \boldsymbol{b}'\boldsymbol{x}_q}{\sqrt{\boldsymbol{x}_q'\boldsymbol{\Sigma}\boldsymbol{x}_q + 1}}\right) - \Phi\left(\frac{\psi_q^{j-1} - \boldsymbol{b}'\boldsymbol{x}_q}{\sqrt{\boldsymbol{x}_q'\boldsymbol{\Sigma}\boldsymbol{x}_q + 1}}\right) \end{aligned} \quad (5)$$

The unconditional probability of ordinal outcome j is obtained by integrating the random components $\boldsymbol{\gamma}_{k,q}$ as

$$P(y_q = j) = \int_{\theta_q} \left[\Phi\left(\frac{\psi_q^j - \boldsymbol{b}'\boldsymbol{x}_q}{\sqrt{\boldsymbol{x}_q'\boldsymbol{\Sigma}\boldsymbol{x}_q + 1}}\right) - \Phi\left(\frac{\psi_q^{j-1} - \boldsymbol{b}'\boldsymbol{x}_q}{\sqrt{\boldsymbol{x}_q'\boldsymbol{\Sigma}\boldsymbol{x}_q + 1}}\right)\right] f(\boldsymbol{\theta}_q) d\boldsymbol{\theta}_q \quad (6)$$

where $f(\boldsymbol{\theta}_q)$ is the multivariate normal probability density function of $\boldsymbol{\theta}_q \sim N(c, \Xi)$ and Ξ is a

$\left(\sum_{k=1}^{J-1} L_k\right) \times \left(\sum_{k=1}^{J-1} L_k\right)$ block diagonal matrix with $\boldsymbol{\Omega}_k$ as the kth diagonal matrix.

Modified Likelihood Function to Handle Misclassification

Let $\alpha_{s,t}$ denote the probability that ordinal alternative s is misclassified as ordinal alternative t. Any given alternative s can be classified as one of the J alternatives, so $\sum_{t=1}^{J} \alpha_{s,t} = 1$. Now, if j is the observed ordinal outcome, then the true latent response can be any of the J alternatives. So, the probability of observed ordinal outcome j under misclassification is given by

$$\tilde{P}_q(y_q = j) = \sum_{t=1}^{J} \alpha_{t,j} \times P(y_q = t) \quad (7)$$

In the current empirical context with four injury severity alternatives, the misclassification matrix is given by

$$\begin{bmatrix} \text{True} \downarrow \,||\text{Observed} & j=1 & j=2 & j=3 & j=4 \\ j=1, \text{No Injury} & \alpha_{1,1} & \alpha_{1,2} & \alpha_{1,3} & \alpha_{1,4} \\ j=2, \text{Possible Injury} & \alpha_{2,1} & \alpha_{2,2} & \alpha_{2,3} & \alpha_{2,4} \\ j=3, \text{Non}-\text{Incapacitating Injury} & \alpha_{3,1} & \alpha_{3,2} & \alpha_{3,3} & \alpha_{3,4} \\ j=4, \text{Incapacitating Injury} & \alpha_{4,1} & \alpha_{4,2} & \alpha_{4,3} & \alpha_{4,4} \end{bmatrix} \quad (8)$$

The diagonal elements in the above matrix indicate the probability that the observed and the true response variables are the same, or the probability of correct classification. Any observed ordinal outcome s in the crash database may be because of misclassification (i.e., the true severity outcome was some other alternative t but was misclassified as s) or because of correct classification. The intuitive meaning of the sufficiency conditions for consistency is that probability of observed data being correct must be larger than probability of being misclassified. If these sufficiency conditions fail, the parameter estimates in the model can have opposite signs from a model that ignores misclassification and there is little hope in recovering the true parameters consistently (*17*). Mathematically, the sufficiency condition translates into the following equation for alternative s:

$$\sum_{\substack{t=1 \\ t \neq s}}^{J} \alpha_{t,s} < \alpha_{s,s} \forall s \in [1, J] \quad (9)$$

Adding $\sum_{\substack{t=1 \\ t \neq s}}^{J} \alpha_{s,t}$ to both sides of Equation 9 gives

$$\sum_{\substack{t=1 \\ t \neq s}}^{J} \alpha_{t,s} + \sum_{\substack{t=1 \\ t \neq s}}^{J} \alpha_{s,t} < \alpha_{s,s} + \sum_{\substack{t=1 \\ t \neq s}}^{J} \alpha_{s,t} \forall s \in [1, J] \quad (10)$$

But, $\alpha_{s,s} + \sum_{\substack{t=1 \\ t \neq s}}^{J} \alpha_{s,t} = \sum_{t=1}^{J} \alpha_{s,t} = 1$. So, the sufficiency condition is equivalent to

$$\sum_{\substack{t=1 \\ t \neq s}}^{J} \alpha_{t,s} + \sum_{\substack{t=1 \\ t \neq s}}^{J} \alpha_{s,t} < 1 \forall s \in [1, J] \quad (11)$$

For the current empirical application, these sufficiency conditions are

$$(\alpha_{2,1} + \alpha_{3,1} + \alpha_{4,1}) + (\alpha_{1,2} + \alpha_{1,3} + \alpha_{1,4}) < 1 \quad (12a)$$

$$(\alpha_{1,2} + \alpha_{3,2} + \alpha_{4,2}) + (\alpha_{2,1} + \alpha_{2,3} + \alpha_{2,4}) < 1 \quad (12b)$$

$$(\alpha_{1,3} + \alpha_{2,3} + \alpha_{4,3}) + (\alpha_{3,1} + \alpha_{3,2} + \alpha_{3,4}) < 1 \quad (12c)$$

$$(\alpha_{1,4} + \alpha_{2,4} + \alpha_{3,4}) + (\alpha_{4,1} + \alpha_{4,2} + \alpha_{4,3}) < 1 \quad (12d)$$

The probability of the observed outcome in Equation 7 is a function of all the misclassification rates $\alpha_{t,j}$ (please note that the observed severity outcome j varies across crashes q) in addition to all the parameters in the propensity and threshold components of $P(y_q = t)$. So, the analyst must estimate $J \times (J-1)$ additional parameters to account for misclassification. However, in the current empirical context, all the alternatives are ordered. So, it is expected that the misclassification probability $\alpha_{s,t}$ decreases considerably as $|s - t|$ increases. So, several entries of the misclassification matrix are expected to be zero.

The parameters of the MGORP model were estimated using the maximum simulated likelihood (MSL) estimation approach using 150 Halton draws.

Empirical Application

The data used for undertaking the analysis was obtained from the 2014 GES database maintained by the National Highway Traffic Safety Administration (NHTSA)'s National Center for Statistics and Analysis. The GES database is a nationally representative sample of police-recorded accidents that involved at least one motor vehicle traveling on a traffic-way and resulted in property damage, injury, or death. The database provided detailed information of about 53,000 accidents involving 93,000 vehicles including (*a*) details of all people involved in the crash (age, gender, seating position, seat-belt use, alcohol involvement, whether the occupant was ejected, and injury severity level sustained), and (*b*) attributes of all vehicles involved in the crash (body type of the vehicle and whether the vehicle rolled over), roadway geometry (details of the regulatory signs/control at the accident location, number of lanes, roadway type, and speed limit), environmental factors (lighting and weather conditions), and crash characteristics (type of collision, whether the collision occurred at an intersection, and number of vehicles involved). The injury severity of each

Table 1. Misclassification Probabilities in MMGORP Model

True ↓ ‖Observed →	No injury	Possible injury	Non-incapacitating injury	Incapacitating or fatal injury
No injury	1.0000	0.0000	0.0000	0.0000
Possible injury	0.3177	0.6823	0.0000	0.0000
Non-incapacitating injury	0.0000	0.2980	0.6275	0.0745
Incapacitating or fatal injury	0.0000	0.0000	0.0000	1.0000

occupant was recorded on a five-point KABCO ordinal scale: (1) no injury; (2) possible injury; (3) non-incapacitating injury; (4) incapacitating injury; and (5) fatal injury. Given that the focus of this analysis is on accidents involving colliding motor-vehicles, all noncollision crash records, motor-cycle crashes, and crashes involving bicyclists and pedestrians were excluded from the analysis. Additionally, the analysis was limited to driver injury severity. So, all records corresponding to passengers were removed. Lastly, after cleaning the data and eliminating the records with missing information on key explanatory variables and the injury severity variable, the size of the data reduced to about 42,100 driver records from 25,708 crashes. In this final estimation sample, the percentage of drivers who sustained fatal injury was less than 1%. Because of this low percentage of the fatal injury records, the fatal and incapacitating injury categories were combined and labeled as "incapacitating or fatal" injury. The distribution of the dependent variable in the final estimation sample was no injury (69%), possible injury (12.3%), non-incapacitating injury (12.2%), and incapacitating or fatal injury (6.5%).

Past research findings, statistical significance, and parameter intuitiveness guided the model estimation. Only parameters that were statistically significant at 95% confidence level were retained in the final model specification. Two models—misclassification-adjusted MGORP (MMGORP) and standard MGORP model that ignores misclassification—were estimated for comparison purposes.

Misclassification Rates

Table 1 presents the estimated misclassification matrix in the best specification of the MMGORP model. The identification conditions in Equations 12*a*–12*d* are satisfied by the estimated misclassification rates. Significant non-zero misclassification rates were found for three pairs of injury severity outcomes. Specifically, 31.77% of possible injuries were wrongly recorded as no injuries; 29.8% of non-incapacitating injuries were wrongly classified as possible injuries; and 7.45% of non-incapacitating injuries were wrongly recorded as incapacitating or fatal injuries. Alternatively, only 68.23% and 62.75% of possible and non-capacitating injuries were correctly classified. This is consistent with the expectation that there may be considerable subjectivity while classifying crashes with intermediate severity outcomes (possible and non-incapacitating injuries).

Parameters Interpretation

Table 2 presents the parameter estimates of the MMGORP model and the bias between the parameter estimates of MMGORP (β_{MMGORP}) and MGORP (β_{MGORP}) models computed as

$$Bias = \frac{(\beta_{MGORP} - \beta_{MMGORP})}{\beta_{MMGORP}} \times 100 \tag{13}$$

All the parameter estimates have the same sign in the two models except for the constant parameter in the last threshold. However, there was considerable bias in the parameter estimates of the MGORP model that ignores misclassification as shown in the last column of Table 2. The MGORP model seems to underestimate the parameters in the propensity and second threshold and overestimate the parameters in the third threshold relative to the MMGORP model. This finding is consistent with the over-representation of the less severe possible injury category because of misclassification in the more severe non-incapacitating injury category. So, the MGORP model skews the propensity to the left and the second and third thresholds to the right to account for over-representation of the possible injury category and under-representation of the non-incapacitating injury category. Additionally, the bias in the threshold parameters was higher than bias in the propensity parameters. Specifically, the average absolute percentage bias values were 13.2%, 27.3%, 38.4%, and 23.8% for the propensity, first, second, and third threshold parameters, respectively. So, the bias in the second threshold parameters between possible and non-incapacitating injury categories and in the third threshold parameters between incapacitating and non-incapacitating injury categories was higher compared with other parameters. This is consistent with significant misclassification rate in the non-incapacitating injury category. However, the bias, although relatively lower in magnitude, propagates to the propensity and other threshold parameters.

From an interpretation standpoint, higher propensity values translate into higher probabilities of more severe injury outcomes. What is more, higher values of the

Table 2. The MMGORP Model Results

Explanatory variables	Propensity		Second threshold		Third threshold	
	Coeff.	Bias	Coeff.	Bias	Coeff.	Bias
Constant			−1.095	30.20	0.438	125.44
Standard deviation			0.309	7.93	0.333	15.06
Gender (base: female)						
Male	−0.349	12.70	−0.204	5.53	−0.234	15.24
Age (base: <=15 years)						
16–19 years	−0.311	10.63				
20–25 years	−0.128	11.92				
46–60 years	0.122	13.49				
61–75 years	0.128	6.93			−0.124	24.37
>=76 years					−0.192	0.62
DUI (base: no)						
Yes	0.321	7.81				
Wearing a seat belt (base: yes)						
No	1.381	12.05				
Ejected (base: no)						
Yes	1.854	22.61				
Body type (base: passenger car)						
SUV	−0.201	8.76				
Van	−0.153	11.36				
Light truck	−0.273	8.25				
Heavy truck	−1.243	6.82				
Standard deviation	0.717	12.14				
Vehicle rolled over (base: no)						
Yes	1.562	31.53			0.689	25.19
Standard deviation	1.152	28.73				
Roadway type (base: one-way or two-way undivided)						
Two-way divided unprotected	0.234	15.28				
Two-way divided protected	0.254	6.45	0.240	42.99		
Ramp	0.221	2.22				
# of lanes (base: one lane)						
Two lanes	0.222	27.09				
Three lanes	0.209	31.85				
Four or more lanes	0.243	27.88				
Surface condition (base: normal)						
Wet					0.112	29.46
Snow	−0.245	7.42				
Standard deviation	0.635	7.21				
Ice	−0.327	7.80				
Standard deviation	0.528	4.32				
Traffic control (base: no control)						
Traffic signal	−0.077	5.45				
Stop sign	−0.321	3.09				
Standard deviation	0.505	1.23				
Yield sign	−0.545	18.32				
Type of collision						
Front end collision	1.112	13.74	0.377	50.60	−0.313	0.57
Rear end collision	−0.465	4.86				
Angled collision	0.462	13.13	0.264	38.09	−0.181	26.59
Side swipe same direction	−0.426	9.63				
Side swipe opposite direction	0.307	11.45			−0.288	19.86
Crash occurred at intersection (base: no)						
Yes	0.080	13.39				
Lighting conditions (base: day/dark lighting)						
Dark no lighting			−0.578	66.98		
Dawn	0.169	4.87				
Cloudy	0.040	37.06	−0.199	58.59	0.061	1.64
Fixed object crash (base: no)						
Yes	0.743	12.06			−0.249	1.33

(continued)

Table 2. (continued)

Explanatory variables	Propensity		Second threshold		Third threshold	
	Coeff.	Bias	Coeff.	Bias	Coeff.	Bias
Number of vehicles (base: <3)						
3 or more	0.193	30.69	0.310	45.07		
First threshold						
Constant			−0.375 (% Difference = 22.30%)			
Standard deviation			−1.031 (% Difference = 32.35%)			
Log-likelihood at convergence						
MGORP model			−36,410.74			
MMGORP model			−36,352.65			

Note: Coeff. = coefficient; DUI = driving under the influence.

second threshold will lead to higher probability for possible injury and lower probability for non-incapacitating injury. Similarly, higher values of the third threshold will lead to higher probability of non-incapacitating injury and lower probability of incapacitating or fatal injury outcome. Everything else being the same, men tend to sustain less severe injuries compared with women. Older drivers are more likely to sustain severe injuries compared with younger drivers. As expected, driving under the influence of alcohol and driver ejection increase whereas seat-belt use lowers the propensity to sustain severe injuries. Drivers in SUVs, vans, light trucks, and heavy trucks have lower risk propensity compared with passenger car drivers. Drivers in rolled-over vehicles tend to sustain severe injuries. Crashes along two-way divided roadways and ramps tend to be more severe compared with those occurring on one-way and two-way undivided roadways. Drivers in crashes along multilane roadways have higher risk propensity than crashes along single lane roads. Interestingly, speed limit was not found to influence the risk propensity of drivers. This is probably because of correlation between roadway geometry variables (roadway type and number of lanes) and speed limit. Intersection crashes have marginally higher risk propensity compared with crashes elsewhere. The type of traffic control at the intersection also had significant impact on severity. Specifically, accidents at controlled intersections including traffic signals, stop, and yield signs have higher risk propensity compared with crashes at uncontrolled intersections. There is no base category for collision type variable because the constant in the propensity was completely segmented by the type of crash. Front–front, angled, and opposite direction sideswipe collisions tend to be more severe compared with front–rear and same direction sideswipe collisions. Drivers involved in crashes with fixed objects have higher risk propensity compared with drivers involved in vehicular collisions. Dawn and cloudy conditions have marginally higher risk propensity compared with dark and day lighting conditions. The higher the number of vehicles involved in a crash, the higher is the risk propensity. Significant unobserved heterogeneity was observed in the effect of vehicle body type, vehicle rollover indicator, surface condition, and traffic control type as indicated by the standard deviation parameters in Table 2. Additionally, all three thresholds were found to be stochastic as indicated by the standard deviation parameters on the constants in these thresholds. These results underscore the importance of unobserved heterogeneity in crash severity modeling.

Statistical Fit Comparison

The final log-likelihood values of the MGORP and MMGORP models were −36, 410.74 and −36, 352.65, respectively. The MMGORP model that handles misclassification has three additional parameters and nests the MGORP model as a special case. So, these two models can be compared using the log-likelihood ratio (LR) test. The LR statistic of comparison between the two models was equal to $-2 \times (-36352.65 + 36410.74) = 116.17$ which is considerably greater than the critical chi-squared value of 7.81 corresponding to three degrees of freedom. This suggests superior data fit in the MMGORP model. The Akaike Information Criterion ($AIC = -2 \times LL + K \times 2$) and Bayesian Information Criterion ($BIC = -2 \times LL + K \times LN(N)$) statistics were [72,837.30, 72408.05] for the MMGORP model and [72947.48, 73492.28] for the MGORP model, respectively. This suggests that the model with misclassification performs better than the model without adjustments for misclassification because a model with lower AIC/BIC values is preferred.

Elasticity Analysis

The bias in a parameter estimate of a variable does not necessarily mean significantly different policy implications. This is because the parameter estimates in Table 2 do not directly indicate the magnitude of variable effects. To gain an understanding of the relative effects of different variables and the policy implications of misclassification, aggregate elasticity effects were calculated as the percentage change in the shares of different injury severity levels for a unit change in an explanatory variable. First, the shares of different injury severity categories were computed in the base scenario by summing the probabilities obtained using Equation 6 across all observations. Next, the shares were recomputed in the policy scenario using the same equation but with a unit increase in the variable for which the elasticity is being calculated. Given that all variables in the final model specification are indicator variables, the unit change is 0 to 1. Table 3 presents the results of the elasticity analysis for the MMGORP model and the percentage bias in the MGORP model elasticity effects. The elasticity corresponding to the "gender" variable for the "incapacitating or fatal" injury was −5.8% indicating that male drivers are 5.8% less likely to sustain "incapacitating or fatal" injury compared with women in the event of a crash. The next set of four numbers in the first row indicate the bias of elasticity effects in the MGORP model that ignores misclassification. For instance, the MGORP model overestimates the elasticity effect for "incapacitating or fatal" injury by 125%. Other numbers in the table can be interpreted similarly. From the relative magnitude of elasticity effects, it can be seen that ejection from vehicle, front–front vehicle collisions, vehicle rollover, and fixed object collisions are the most likely scenarios to result in "incapacitating or fatal" injury. Similarly, seat-belt use, drivers in heavy trucks, rear–rear collisions, yield sign traffic control, and younger drivers are least likely scenarios to result in "incapacitating or fatal" injury. The results indicate considerable bias in the elasticity estimates of the MGORP model. Furthermore, the bias is much higher for the "possible injury" and "non-incapacitating injury" categories compared with "no injury" and "incapacitating or fatal injury" categories. For instance, drivers under the influence of alcohol are 6.2% more likely to sustain possible injury. However, the MGORP model overestimates this elasticity effect by 111.5%. In fact, the average absolute bias values for the four injury severity levels are 8.5%, 64.1%, 36.7%, and 15.7%, respectively. These findings are consistent with the statistically significant misclassification rates corresponding to the possible and non-incapacitating injury categories in Table 1. Overall, the results indicate that misclassification in injury severity data can result in biased parameter estimates leading to incorrect policy sensitivity results.

Conclusions

The police-reported injury severity recordings in crash databases are prone to errors. Past research that measured the discordance between police-reported injury severity data and hospital/ambulance records confirmed the presence of misclassification errors in traditional crash databases. However, these databases remain the primary data sources for safety analysis including aggregate crash frequency and disaggregate injury severity analysis conditional on crash occurrence. Ignoring the errors in the injury severity data during modeling can lead to biased and inconsistent parameter estimates. It is surprising that none of the earlier studies attempted to quantify and adjust the bias caused by misclassification in injury severity models. In this study, a misclassification-adjusted mixed generalized ordered response probit (MMGORP) model was developed to analyze driver injury severity using the 2014 GES data. The research methodology adopted in this paper treats the observed injury severity outcomes as realizations from discrete random variables that depend on true latent injury severities that are unobservable to the analyst.

The results indicate that 31.77% of possible injuries were wrongly recorded as no injuries; 29.80% of non-incapacitating injuries were wrongly classified as possible injuries; and 7.45% of non-incapacitating injuries were wrongly recorded as incapacitating or fatal injuries. Additionally, the MGORP model that ignores misclassification not only has lower data fit but also considerable bias in the parameter and elasticity effects leading to incorrect policy implications. The model developed in this study can be used to investigate misclassification errors in ordinal response variables in other empirical contexts beyond transportation safety.

There are several possible avenues for future research. For instance, the results indicate that there is no significant misclassification in the incapacitating/fatal injury category, in other words that all true incapacitating and fatal injuries are recorded correctly. However, it is possible that the misclassification rates in incapacitating injuries are zero because of merging the incapacitating and fatal injury categories as one alternative. This is because fatal crashes are rarely misclassified and the nonzero misclassification in incapacitating injuries are weighed down by the zero misclassification rate in fatal crashes. The impact of aggregation of injury severity categories on misclassification rates is an avenue for future research. Moreover, the misclassification rates in the model developed do not vary across observations. However, earlier studies found that the discordance rates between police and hospital records vary as a function of different factors including driver, crash, and geographic factors. The model developed in this study can be extended to allow the misclassification rates to vary across different segments. For instance, two sets of misclassification rates can be estimated separately for crashes that occur in

Table 3. Elasticity Effects of MMGORP Model

Explanatory variables/alternative	GORP- misclassification				% Bias of MGORP model			
	NI	PI	NII	IFI	NI	PI	NII	IFI
Gender (base: female)								
Male	12.2	−25.6	−29.4	−5.8	−6.7	9.0	−1.5	125.0
Age (base: less than <=15 years)								
16–19 years	9.8	−10.7	−22.6	−32.3	−5.4	39.5	5.6	−2.5
20–25 years	4.1	−3.9	−9.5	−14.6	−5.8	50.0	5.7	−4.3
46–60 years	−3.9	3.2	9.4	15.8	−6.5	63.1	5.5	−6.8
61–75 years	−4.2	3.3	1.6	33.7	0.9	79.1	127.0	−26.7
>=76 years	−6.8	4.7	1.9	56.0	1.2	91.1	−2.2	−20.0
DUI (base: no)								
Yes	−10.7	6.2	24.2	46.5	1.5	111.5	15.5	−0.6
Wearing a seat belt (base: yes)								
No	−45.2	−12.7	72.0	325.8	4.6	−233.9	32.0	−8.1
Ejected (base: no)								
Yes	−58.1	−30.6	63.6	453.0	−4.5	−127.4	49.1	−22.6
Vehicle type (base: passenger car)								
SUV	6.5	−6.2	−14.6	−22.0	−2.7	52.2	8.9	−1.0
Van	4.9	−4.9	−11.4	−17.3	−5.5	46.8	5.9	−3.6
Light truck	8.7	−9.0	−19.8	−28.8	−2.8	47.2	8.7	−0.4
Heavy truck	28.1	−53.0	−64.3	−69.7	−5.6	10.2	4.7	3.2
Vehicle rolled over (base: no)								
Yes	−45.9	−26.8	151.0	179.4	−13.1	−108.9	2.9	2.5
Roadway type (base: one-way or two-way undivided)								
Two-way divided unprotected roadway	−7.6	33.4	11.2	18.8	−8.0	−27.9	13.5	1.8
Two-way divided protected roadway	−8.2	6.3	20.1	35.7	1.6	86.0	15.3	1.3
Ramp	−7.3	4.9	16.6	30.0	7.1	103.5	21.2	6.3
Number of lanes (base: one lane)								
Two lanes	−7.0	6.0	17.8	31.0	−20.8	38.2	−11.7	−23.3
Three lanes	−6.8	5.3	16.2	28.5	−26.1	33.7	−17.1	−28.4
Four or more lanes	−7.7	6.6	19.4	33.8	−21.6	35.7	−12.8	−24.1
Surface conditions (base: normal)								
Snow	5.7	−15.6	−15.3	−5.9	5.0	12.2	12.5	73.6
Ice	8.5	−16.3	−21.1	−20.2	−9.6	13.1	0.5	−3.6
Traffic control type (base: no control)								
Traffic signal	2.5	−2.2	−5.7	−9.0	1.5	66.4	13.9	2.3
Stop sign	8.6	−15.7	−20.9	−20.8	−5.2	20.2	5.3	0.1
Yield sign	16.2	−21.9	−38.6	−50.3	−13.9	14.6	−4.0	−8.2
Type of collision								
Front end collision	−36.8	48.0	28.6	273.7	0.1	5.6	31.6	−15.6
Rear end collision	14.0	−18.1	−33.5	−44.7	−1.1	37.9	10.3	2.7
Angled collision	−14.7	44.3	16.7	86.1	−5.4	−10.8	25.3	−23.4
Side swipe same direction	13.3	−15.9	−30.7	−41.9	−5.3	34.1	5.7	−1.4
Side swipe opposite direction	−10.2	6.0	0.7	88.6	−2.8	101.4	−733.5	−18.0
Crash occurred at intersection (base: no)								
Yes	−2.6	2.3	6.2	10.2	−6.8	55.9	4.9	−6.6
Lighting conditions (base: day)								
Dawn	−5.5	4.0	12.7	22.3	14.6	108.7	29.4	14.6
Cloudy	−1.3	−16.5	11.9	5.9	47.9	−72.5	12.4	28.9
Fixed object collision (base: no)								
Yes	−24.4	6.9	29.9	213.8	−1.2	253.0	14.3	−19.3
Number of vehicles involved (base: <3 vehicles)								
3 or more	−6.3	41.2	5.8	8.9	−25.0	−41.6	−14.1	−19.7

Note: NI = no injury; PI = possible injury; NII = non-incapacitating injury; IFI = incapacitating or fatal injury; DUI = driving under the influence.

urban and rural neighborhoods. The sufficiency conditions in Equation 11 must hold within urban and rural neighborhoods separately (but not necessarily in the two regions together). However, the number of misclassification parameters can explode easily as the number of segments increases. To avoid this problem, latent class models that probabilistically assign each driver/crash record to latent segments each with its own set of misclassification rates can be developed (*28, 29*). The probability of belonging to each latent segment can be specified as a function of driver, vehicle, and crash variables. Recently, this latent modeling approach was used for analyzing misclassification rates in household auto-ownership responses in travel surveys (*30*). Next, not only injury severity recordings but also other variables in crash databases are prone to misclassification. For instance, police have the tendency to overestimate seat-belt use in road casualties (*31*). Seat-belt use also has the endogeneity problem whereby there can be common unobserved factors that influence the decision to wear a seat belt and the injury severity outcome (*32*). In this context, future research that develops an integrated modeling framework to account for misclassification in key explanatory variables in addition to the injury severity response variable is warranted.

Author Contributions

The authors confirm contribution to the paper as follows: study conception and design: R. Paleti; data collection: L. Balan; analysis and interpretation of results: L. Balan, R. Paleti; draft manuscript preparation: L. Balan. R. Paleti. All authors reviewed the results and approved the final version of the manuscript.

References

1. World Health Organization. *Global Status Report on Road Safety 2015*. World Health Organization, Geneva, Switzerland, 2015.
2. National Highway Traffic Safety Administration. 2015 Motor Vehicle Crashes: Overview. *Traffic Safety Facts Research Note*, Vol. 2016, 2016, pp. 1–9.
3. Tsui, K., F. So, N.-N. Sze, S. Wong, and T.-F. Leung. Misclassification of Injury Severity among Road Casualties in Police Reports. *Accident Analysis & Prevention*, Vol. 41, No. 1, 2009, pp. 84–89.
4. Farmer, C. M. Reliability of Police-Reported Information for Determining Crash and Injury Severity. *Traffic Injury Prevention*, Vol. 4, 2003, pp. 38–44.
5. O'Day, J. *NCHRP Synthesis of Highway Practice 192: Accident Data Quality*. HRB, National Research Council, Washington, D.C., 1993.
6. Burch, C., L. Cook, and P. Dischinger. A Comparison of KABCO and AIS Injury Severity Metrics Using CODES Linked Data. *Traffic Injury Prevention*, Vol. 15, No. 6, 2014, pp. 627–630.
7. Alsop, J., and J. Langley. Under-Reporting of Motor Vehicle Traffic Crash Victims in New Zealand. *Accident Analysis & Prevention*, Vol. 33, No. 3, 2001, pp. 353–359.
8. Sciortino, S., M. Vassar, M. Radetsky, and M. M. Knudson. San Francisco Pedestrian Injury Surveillance: Mapping, Under-Reporting, and Injury Severity in Police and Hospital Records. *Accident Analysis & Prevention*, Vol. 37, No. 6, 2005, pp. 1102–1113.
9. Amoros, E., J.-L. Martin, M. Chiron, and B. Laumon. Road Crash Casualties: Characteristics of Police Injury Severity Misclassification. *Journal of Trauma and Acute Care Surgery*, Vol. 62, No. 2, 2007, pp. 482–490.
10. Loo, B. P., and K. Tsui. Factors Affecting the Likelihood of Reporting Road Crashes Resulting in Medical Treatment to the Police. *Injury Prevention*, Vol. 13, No. 3, 2007, pp. 186–189.
11. McDonald, G., G. Davie, and J. Langley. Validity of Police-Reported Information on Injury Severity for Those Hospitalized from Motor Vehicle Traffic Crashes. *Traffic Injury Prevention*, Vol. 10, No. 2, 2009, pp. 184–190.
12. Watson, A., B. Watson, and K. Vallmuur. Estimating Under-Reporting of Road Crash Injuries to Police Using Multiple Linked Data Collections. *Accident Analysis & Prevention*, Vol. 83, 2015, pp. 18–25.
13. Couto, A., M. Amorim, and S. Ferreira. Reporting Road Victims: Assessing and Correcting Data Issues through Distinct Injury Scales. *Journal of Safety Research*, Vol. 57, 2016, pp. 39–45.
14. Aptel, I., L. R. Salmi, F. Masson, A. Bourdé, G. Henrion, and P. Erny. Road Accident Statistics: Discrepancies between Police and Hospital Data in a French Island. *Accident Analysis & Prevention*, Vol. 31, No. 1, 1999, pp. 101–108.
15. Lopez, D. G., D. L. Rosman, G. A. Jelinek, G. J. Wilkes, and P. C. Sprivulis. Complementing Police Road-Crash Records with Trauma Registry Data—an Initial Evaluation. *Accident Analysis & Prevention*, Vol. 32, No. 6, 2000, pp. 771–777.
16. Hausman, J. Mismeasured Variables in Econometric Analysis: Problems from the Right and Problems from the Left. *The Journal of Economic Perspectives*, Vol. 15, No. 4, 2001, pp. 57–67.
17. Hausman, J. A., J. Abrevaya, and F. M. Scott-Morton. Misclassification of the Dependent Variable in a Discrete-Response Setting. *Journal of Econometrics*, Vol. 87, No. 2, 1998, pp. 239–269.
18. Austin, K. The Identification of Mistakes in Road Accident Records: Part 2, Casualty Variables. *Accident Analysis & Prevention*, Vol. 27, No. 2, 1995, pp. 277–282.
19. Osler, T., S. P. Baker, and W. Long. A Modification of the Injury Severity Score That Both Improves Accuracy and Simplifies Scoring. *Journal of Trauma and Acute Care Surgery*, Vol. 43, No. 6, 1997, pp. 922–926.
20. Yamamoto, T., J. Hashiji, and V. N. Shankar. Underreporting in Traffic Accident Data, Bias in Parameters and the Structure of Injury Severity Models. *Accident Analysis & Prevention*, Vol. 40, No. 4, 2008, pp. 1320–1329.
21. Kumara, S., and H. Chin. Application of Poisson Underreporting Model to Examine Crash Frequencies at Signalized Three-Legged Intersections. *Transportation Research Record: Journal of the Transportation Research Board*, 2005. 1908: 46–50.
22. Patil, S., S. R. Geedipally, and D. Lord. Analysis of Crash Severities Using Nested Logit Model—Accounting for the Underreporting of Crashes. *Accident Analysis & Prevention*, Vol. 45, 2012, pp. 646–653.

23. Boes, S., and R. Winkelmann. Ordered Response Models. *Allgemeines Statistisches Archiv*, Vol. 90, No. 1, 2006, pp. 167–181.
24. Greene, W. H., and D. A. Hensher. *Modeling Ordered Choices: A Primer*. Cambridge University Press, Cambridge, United Kingdom, 2010.
25. Eluru, N., C. R. Bhat, and D. A. Hensher. A Mixed Generalized Ordered Response Model for Examining Pedestrian and Bicyclist Injury Severity Level in Traffic Crashes. *Accident Analysis & Prevention*, Vol. 40, No. 3, 2008, pp. 1033–1054.
26. Mannering, F. L., V. Shankar, and C. R. Bhat. Unobserved Heterogeneity and the Statistical Analysis of Highway Accident Data. *Analytic Methods in Accident Research*, Vol. 11, 2016, pp. 1–16.
27. Fountas, G., and P. C. Anastasopoulos. A Random Thresholds Random Parameters Hierarchical Ordered Probit Analysis of Highway Accident Injury-Severities. *Analytic Methods in Accident Research*, Vol. 15, 2017, pp. 1–16.
28. Dustmann, C., and A. Van Soest. An Analysis of Speaking Fluency of Immigrants Using Ordered Response Models with Classification Errors. *Journal of Business & Economic Statistics*, Vol. 22, No. 3, 2004, pp. 312–321.
29. Sullivan, P. Estimation of an Occupational Choice Model When Occupations are Misclassified. *Journal of Human Resources*, Vol. 44, No. 2, 2009, pp. 495–535.
30. Paleti, R., and L. Balan. *Misclassification in Travel Surveys and Implications to Choice Modeling: Application to Household Auto Ownership Decisions*. Transportation, 2017, pp. 1–19.
31. Schiff, M. A., and P. Cummings. Comparison of Reporting of Seat Belt Use by Police and Crash Investigators: Variation in Agreement by Injury Severity. *Accident Analysis & Prevention*, Vol. 36, No. 6, 2004, pp. 961–965.
32. Eluru, N., and C. R. Bhat. A Joint Econometric Analysis of Seat Belt Use and Crash-Related Injury Severity. *Accident Analysis & Prevention*, Vol. 39, No. 5, 2007, pp. 1037–1049.

The Standing Committee on Statistical Methods (ABJ80) peer-reviewed this paper (18-02227).

Article

TRR
JOURNAL OF THE TRANSPORTATION RESEARCH BOARD

Transportation Research Record
2018, Vol. 2672(30) 64–75
© National Academy of Sciences:
Transportation Research Board 2018
Article reuse guidelines:
sagepub.com/journals-permissions
DOI: 10.1177/0361198118801348
journals.sagepub.com/home/trr
SAGE

A Multivariate Copula-Based Macro-Level Crash Count Model

Shamsunnahar Yasmin[1], Salah Uddin Momtaz[1], Tammam Nashad[1], and Naveen Eluru[1]

Abstract

The current study contributes to safety literature both methodologically and empirically by developing a macro-level multivariate copula-based crash frequency model for crash counts. The multivariate model accommodates for the impact of observed and unobserved effects on zonal level crash counts of different road user groups including car, light truck, van, other motorized vehicle (including truck, bus and other vehicles), and non-motorists (including pedestrians and cyclists). The proposed model is estimated using Statewide Traffic Analysis Zone (STAZ) level road traffic crash data for the state of Florida. A host of variable groups including land-use characteristics, roadway attributes, traffic characteristics, socio-economic characteristics and demographic characteristics are considered. The model estimation results illustrate the applicability of the proposed framework for multivariate crash counts. Model estimation results are further augmented by evaluation of predictive performance and policy analysis.

Road traffic crashes affect society as a whole both emotionally and economically and are rightfully recognized as a national health problem (*1*, *2*). In reducing the undue burden of road crashes and their consequences, road safety literature is focused on devising both proactive and reactive safety management policies at the user, system and/or planning levels through evidence-based and data-driven strategies. Crash frequency analysis, specifically macro-level crash models, is a major component for devising and evaluating these road safety policies at a planning level. Macro-level studies have mostly evolved in safety research to incorporate safety considerations within the transportation planning process. The outcome of these models is also useful for devising safety-conscious decision support tools to facilitate a proactive approach in assessing medium and long-term policy-based countermeasures. The current research effort contributes to the safety literature methodologically and empirically with a specific focus on macro-level crash frequency analysis.

Econometric approaches of developing crash prediction models in safety literature are dominated by traditional count regression frameworks (Poisson and negative binomial (NB) models) in univariate modeling systems (see [*3*–*5*]). These studies identify a single count variable for different crash attribute levels (road user group, crash severity, crash types, or vehicles types) for a spatial unit and study the impact of exogenous variables. However, as documented in literature, crash counts across different attribute levels are likely to be dependent for the same observation resulting in a multivariate crash event set (*6*). Ignoring such correlation, if present, may lead to biased and inefficient parameter estimates resulting in erroneous policy implications (*7*, *8*). To that extent, road safety researchers and analysts have estimated multivariate count models to produce more accurate predictions (see (*9*) for a detailed list of these studies).

It is beyond the scope of this paper to provide a comprehensive literature review on multivariate crash count models. For a detailed review of multivariate frameworks employed in safety, the reader is referred to recent review studies (*3*, *10*, *11*). Within the multivariate scheme, studies have predominantly explored crash counts by severity outcome levels and by crash types. However, multivariate crash event sets may also arise when examining crash occurrences by different road user groups involved in

[1]Department of Civil, Environmental & Construction Engineering, University of Central Florida, Orlando, FL

Corresponding Author:
Address correspondence to Shamsunnahar Yasmin: shamsunnahar.yasmin@ucf.edu

crashes. In fact, studies have recognized this and developed multivariate crash count models for different road user groups involved in crashes – for pedestrians and cyclists (*12*, *13*), for vehicle types (*14*), and for travel modes (*15*).

In these studies, the general trend is to focus entirely either on motorized road user groups or on non-motorized road users (except [*15*]). However, both of these road user groups share the same travel environment within a spatial planning unit over a specific given period of time. Therefore, it is possible that the same set of observed and unobserved factors influence crash occurrences of these two different road user groups. For instance, higher numbers of uncontrolled intersections (usually observed by analysts) at a zonal level are likely to result in higher numbers of vehicular conflicts as well as higher numbers of crashes involving pedestrians or cyclist. At the same time, if a zone has a higher proportion of blind spots at intersections (usually unobserved by analysts) it may contribute to more crash events involving both motorists and non-motorists. Therefore, it is important to examine crash events as joint processes considering both of these road user groups simultaneously. Further, while analyzing motorized road user groups, recognizing the implicit differences between various motorized vehicle groups is very useful. It is plausible that different exogenous variables may have distinct impacts on crash occurrence across various motorized road user groups. For instance, zones with higher truck volumes may have higher numbers of crashes involving heavy vehicles. Moreover, it is also important to examine separate risk factors related to different types of passenger vehicles rather than considering all passenger vehicles as one category. As documented in literature, the diversity in the passenger vehicle fleet has deteriorating effects on overall safety (*16*). In the United States, sales of light truck increased 7% in 2016 relative to 2015 (*17*). The shift from light to heavy passenger vehicles is likely to result in 4.3 additional crashes (for each fatal crash that occupants of large passenger vehicles avoid) that may result in fatalities among occupants of light vehicles or non-motorists involved in crashes with these heavy passenger vehicles (*18*).

Given the potential difference in the safety impacts of different types of passenger vehicles, it is important to examine separate risk factors for different types of passenger vehicles, which would allow us to devise more tangible actions and policies. The first contribution of our study is to develop a multivariate crash count model for crashes involving different road user groups involved in crashes with higher resolution classification of the passenger vehicle fleet. Specifically, we examine zonal level car, light truck, van, other motorized vehicles (bus, truck and other vehicles), and non-motorist (pedestrians and cyclists) involved crash counts in a multivariate count model framework.

Traditionally, in existing safety literature, the multivariate count models are examined by considering unobserved error components that jointly affect the dependent variables. In particular, the traditional multivariate count modeling approaches partition the error components of the dependent variables to accommodate for a common term and an independent term across dependent variables (see (*6*) for a detailed discussion of various methodologies). Thus, any probability computation, in accommodating such unobserved effects, requires integrating the probability function over the error term distribution. The exact computation is dependent on the distributional assumption and does not usually have a closed-form expression. Thus, the estimation procedure requires the adoption of maximum simulated likelihood (MSL) approach in the classical approach or Markov Chain Monte Carlo (MCMC) approach in the Bayesian realm. MSL and MCMC methods provide substantial flexibility in accommodating unobserved heterogeneity. However, the probability evaluation with high dimensional integrals is affected by the challenges in generating high dimensionality of random numbers and longer computational run times. The process of applying simulation for such joint processes is likely to be error-prone and the stability of the variance-covariance matrix is often sensitive to model specification and a number of simulation draws (see (*19*) for a discussion). Within this simulated framework, the model structures employed in developing multivariate crash count models include multivariate-Poisson, multivariate Poisson-lognormal, multivariate random-parameters zero-inflated NB, multinomial-generalized Poisson, multivariate conditional autoregressive, multivariate tobit and multivariate Poisson gamma mixture count models. Another multivariate count modeling approach based on the development of multivariate functions has most recently been employed by Narayanamoorthy et al. (*20*). The approach circumvents the challenges associated with simulation by adopting analytical approximation of the likelihood function.

More recently, a closed-form parametric formulation that obviates the need for an approximation or demanding simulation has been employed in existing econometric literature for examining joint count events. The approach, referred to as the copula-based approach, allows for flexible correlation structures across joint dimensions thus enhancing the flexibility of the multivariate approach. The copula-based approach allows for analytical computation of the log-likelihood based on standard maximum likelihood procedures; it is generally tractable and offers a stable inference. The copula formulation allows for additional flexibility in specifying the marginal distribution. While the application of copula has seen a surge of interest in examining multivariate continuous and disaggregate discrete data, the studies employing copulas for examining aggregate level count events are relatively few (for application of copulas in continuous and disaggregate level discrete data see [*21–24*]). The copula-based bivariate count model has been employed in econometrics and applied statistics (*25*, *26*). To date, only one study in safety literature has employed a bivariate copula-count model in examining pedestrian and bicycle crash risks simultaneously (*12*).

The current study generalizes the bivariate copula-count model for examining multivariate count data. Specifically, we formulate and estimate a multivariate copula-count model for examining zonal level crash counts by different road user groups involved in crashes. To be sure, the application of multivariate copula-count models has been demonstrated by Nikoloulopoulos and Karlis (*27*) in examining the correlation among the number of purchases of four different products (food, non-food, hygiene, and fresh). In context of the current study, we employ multivariate copula-count models to examine five different crash count dimensions: car, light truck, van, other motorized vehicle and crashes involving non-motorists. The second contribution of our study is to develop a closed-form multivariate copula-count model to accommodate for the impact of observed and unobserved effects on zonal level crash counts of different road user groups. To examine the count components of the multivariate copula-based model, we employ NB regression frameworks. The NB model that has a built-in dispersion parameter is widely employed in safety literature. It provides a natural enhancement over the Poisson model and is easy to estimate with a closed-form structure to accommodate for overdispersion (the variance of the crash count variable usually exceeds the mean of the crash count variable). In existing safety literature, researchers have also employed count modeling frameworks accommodating the preponderance of zero count events (such as zero-inflated and hurdle models). However, NB is the most frequently used statistical technique for examining crash count events (*10*). Therefore, in our current study, we examine crash count within the proposed multivariate copula-based approach by using an NB regression framework. The proposed model is estimated using Statewide Traffic Analysis Zone (STAZ) level road crash data for the state of Florida. A host of variable groups including, land-use characteristics, roadway attributes, traffic characteristics, socio-economic characteristics and demographic characteristics, are considered.

In summary, the current research effort contributes to safety literature on macro-level crash count analysis both methodologically and empirically. In terms of methodology, we formulate and estimate a multivariate copula-based count model framework to analyze the crash count events for different road user groups involved in crashes jointly, and we employ NB regression framework for examining the count components. The proposed multivariate copula-count model can be employed in developing both macro and micro-level count events. In terms of empirical analysis, our study incorporates crash counts for both motorized and non-motorized road user groups while considering different types of passenger vehicle fleet categories. Specifically, we examine crash counts for car, light truck, van, other motorized vehicle, and crashes involving non-motorists by employing a multivariate copula-count framework. Model estimation results are further augmented by evaluation of predictive performance and policy analysis.

Econometric Framework

The focus of our study is to propose and estimate a copula-based multivariate NB modeling framework. See (*22*, *28*) for a detailed background on copula-based models and (*27*) for a description of multivariate NB frameworks. The econometric framework for the joint model is presented in this section.

Let us assume that i is the index for STAZ $(i = 1, 2, 3, \ldots, N)$ and y_{qi} the index for crashes occurring over a period of time in a STAZ i; $q(q = 1, 2, \ldots, M, M = 5)$ is the index to represent road user groups for the multivariate case examined. In this empirical study, q takes the value of "car"$(q = 1)$, "light truck"$(q = 2)$, "van"$(q = 3)$, "other motorized vehicle"$(q = 4)$, and "non-motorist"$(q = 5)$. The NB probability expression for random variable y_{qi} can be written as:

$$P_{qi}\left(y_{qi}|\mu_{qi}, \alpha_q\right) = \frac{\Gamma\left(y_{qi} + \alpha_q^{-1}\right)}{\Gamma(y_{qi} + 1)\Gamma\left(\alpha_q^{-1}\right)}\left(\frac{1}{1 + \alpha_q\mu_{qi}}\right)^{\frac{1}{\alpha_q}}\left(1 - \frac{1}{1 + \alpha_q\mu_{qi}}\right)^{y_{qi}} \quad (1)$$

where $\Gamma(\cdot)$ is the Gamma function, α_q is the NB dispersion parameter specific to the road user group q, and μ_{qi} is the expected number of crashes occurring in STAZ i over a given period of time for road user group q. We can express μ_{qi} as a function of explanatory variable $(\boldsymbol{x}_{qi})$ by using a log-link function as: $\mu_{qis} = E\left(y_{qi}|\boldsymbol{x}_{qi}\right) = exp\left(\boldsymbol{\beta}_q\boldsymbol{x}_{qi}\right)$, where $\boldsymbol{\beta}_q$ is a vector of parameters to be estimated specific to road user group q.

The correlation or joint behavior of random variables $y_{1i}, y_{2i}, \ldots y_{Mi}$ is explored in the current study by using a copula-based approach. A copula is a mathematical device that identifies dependency among random variables with pre-specified marginal distribution. (*22*, *29*) provide a detailed description of the copula approach. In constructing the copula dependency, let us assume that $\Lambda_1(y_{1i})$, $\Lambda_2(y_{2i})...\Lambda_M(y_{Mi})$ are the marginal distribution functions of the random variables y_{1i}, y_{2i},...y_{Mi}, respectively; and $\Lambda_{12...M}(y_{1i}, y_{2i}, \ldots y_{Mi})$ is the M variate joint distribution with corresponding marginal distributions. Subsequently, the M variate distribution $\Lambda_{12...M}(y_{1i}, y_{2i}, \ldots y_{Mi})$ can be generated as a joint cumulative probability distribution of uniform [0, 1] marginal variables U_1, U_2 ... U_M as follows:

$$\begin{aligned} &\Lambda_{12...M}(y_{1i}, y_{2i}, \ldots y_{Mi}) \\ &= \Pr(U_1 \leqslant y_{1i}, U_2 \leqslant y_{2i} \ldots, U_M \leqslant y_{Mi}) \\ &= Pr[\Lambda_1^{-1}(U_1) \leqslant y_{1i}, \Lambda_2^{-1}(U_2) \leqslant y_{2i} \ldots, \Lambda_M^{-1}(U_M) \leqslant y_{Mi}] \\ &= Pr[U_1 < \Lambda_1(y_{1i}), U_2 < \Lambda_2(y_{2i}) \ldots, U_M < \Lambda_M(y_{Mi})] \end{aligned} \tag{2}$$

The joint distribution (of uniform marginal variable) in Equation 2 can be generated by a function $C_{\theta i}(.,.)$ (*30*), such that:

$$\begin{aligned} &\Lambda_{12...M}(y_{1i}, y_{2i}, \ldots y_{Mi}) \\ &= C_{\theta i}(U_1 = \Lambda_1(y_{1i}), U_2 = \Lambda_2(y_{2i}) \ldots, U_M = \Lambda_M(y_{Mi})) \end{aligned} \tag{3}$$

where, $C_{\theta i}(.,.)$ is a copula function and θ_i is the dependence parameter defining the link between $y_{1i}, y_{2i}, \ldots y_{Mi}$. In the case of continuous random variables, the joint density can be derived from partial derivatives. However, in our study, y_{qi} are nonnegative integer valued events. For such count data, following (*26*), the probability mass function ($\zeta_{\theta i}$) is presented (instead of continuous derivatives) by using finite differences of the copula representation as follows:

$$\begin{aligned} &\zeta_{\theta i}(\Lambda_1(y_{1i}), \Lambda_2(y_{2i}) \ldots \Lambda_M(y_{Mi})) \\ &= \sum_{a_1=1}^{2} \sum_{a_2=1}^{2} \cdots \sum_{a_M=1}^{2} (-1)^{a_1+a_2+\ldots a_M} [C_{\theta i}(\Lambda_1(y_{1i}+a_1-2), \Lambda_2(y_{2i}+a_2-2) \ldots \Lambda_M(y_{Mi}+a_M-2); \theta_i)] \end{aligned} \tag{4}$$

The reader would note the probability in Equation 4 is written in terms of 2^M copula evaluations (see (*31*, *32*) for a similar derivation). The number of computations increases rapidly with the number of dependent variables (M), but this is not much of a problem when the dependent variable number M is 6 or less because of the closed-form structures of the copula function evaluation. Given the above setup, we specify $\Lambda_1(y_{1i})$, $\Lambda_2(y_{2i})$ ... $\Lambda_M(y_{Mi})$ as the cumulative distribution function (cdf) of the NB distribution. The cdf of NB probability expression (as presented in Equation 1) for y_{qi} can be written as:

$$\Lambda_q\left(y_{qi}|\mu_{qi}, \alpha_q\right) = \sum_{k=0}^{y_{qi}} P_{qi}\left(y_{qi}|\mu_{qi}, \alpha_q\right) \tag{5}$$

Thus, the log-likelihood function (*LL*) with the joint probability expression in Equation 5 can be written as:

$$LL = \sum_{i=1}^{N} \ln(\zeta_{\theta i}(\Lambda_1(y_{1i}), \Lambda_2(y_{2i}) \ldots \Lambda_M(y_{Mi}))) \tag{6}$$

In the current empirical study, we employ Archimedean copulas that span the spectrum of different kinds of dependency structures including Clayton, Gumbel, Frank, and Joe copulas (see (*22*) for graphical descriptions of the implied dependency structures). Archimedean copulas, in their multivariate forms, allow only positive associations and equal dependencies among pairs of random variables. The parameters are estimated using a maximum likelihood approach. The model estimation is achieved through the log-likelihood functions programmed in GAUSS.

Data Description

Our study area includes the state of Florida with 8,518 STAZs. Similar to the other transportation trends seen in North America, the ethos of travel in Florida is also predominantly auto-oriented. The state had nearly 100,000 more crashes in 2015 than in 2011 with a higher number of non-motorist fatalities (*33*). These numbers clearly signify that it is important to identify critical factors contributing to road traffic crashes at a planning level for all road user groups to improve the overall road safety situation. The traffic crash records are collected and compiled from Florida Department of Transportation (FDOT) Crash Analysis Reporting System (CARS) database for the year 2014. The geocoded crash data are aggregated at the level of STAZ for each road user group. Thus, the dependent variable of the empirical study is the zonal level number of crash counts involving car, light truck, van, other motorized vehicles, and non-motorists.

In addition to the crash database, the explanatory attributes considered in the empirical study are also aggregated at the STAZ level. The selected explanatory variables can be grouped into five broad categories: land-use characteristics, roadway attributes, traffic characteristics, socio-economic characteristics, and demographic characteristics. These variables are collected and compiled from different data sources including: 2010 US census data; 2009–2013 American Community Survey (ACS); and Florida Geographic Data Library (FGDL) databases. Land-use characteristics included shopping centers, restaurants, park/recreational centers, and the proportion of the urban area. Roadway attributes included the proportion of local roads and major roads length. Traffic characteristics included annual average daily traffic (AADT) and truck AADT. Socio-economic characteristics included the proportion of industrial and retail jobs, the proportion of households with no vehicle, and the proportion of households with one vehicle. Finally, Demographic characteristics included the proportion of the Hispanic and Caucasian populations.

Table 1 offers a summary of the sample characteristics of the exogenous factors in the estimation dataset along with the descriptive statistics of the dependent variables. The table represents the definition of variables considered for final model estimation along with the zonal minimum, maximum, average values, and standard deviation. The final specification of the model development was based on removing the statistically insignificant variables in a systematic process based on statistical significance (95% significance level). The specification process was also guided by prior research and parsimony considerations. In estimating the models, several functional forms and variable specifications were explored. The functional form that provided the best result was used for the final model specifications and, in Table 1, the variable definitions are presented based on these final functional forms.

Empirical Analysis

Model Selection

The empirical analysis involves estimation of four different multivariate count models: (i) Clayton; (ii) Gumbel; (iii) Frank; and (iv) Joe copulas. We also estimate an independent copula model (separate NB models for crash counts involving different road user groups) to establish a benchmark for comparison. A comparison exercise is undertaken to identify the most suitable copula model (including the independent copula model). The alternative copula models estimated are non-nested and therefore cannot be tested using traditional log-likelihood ratio tests. We employ the Bayesian Information Criterion (BIC) to determine the best model among all copula models (*12*, *21*, *29*, *32*). The model with the lower BIC is the preferred copula model. The BIC value for the independent copula model is 192617.83. The BIC values for the estimated multivariate copula models are: Clayton – 166929.78; Gumbel – 165722.37; Frank – 167869.31; and Joe – 167572.03. From the BIC values, we can see that the estimated copula models provide improved data fit relative to independent models. However, the copula model with Gumbel distribution outperforms all other copula models. The BIC comparisons confirm the importance of accommodating dependence among crash count events of different road user groups in the macro-level analysis.

Estimation Results

In presenting the effects of exogenous variables in the multivariate model specification, we will restrict ourselves to the discussion of the Gumbel copula specification. The estimation results of the multivariate (Gumbel copula) model are presented in Table 2. We include car, light truck, van, other motorized vehicle and non-motorist crash count components in the second, third, fourth, fifth, and sixth column panels, respectively. For brevity, results are discussed together for different road user groups in the following section by variable groups.

Land-Use Characteristics. Among different points of interest considered, the copula model results reveal a higher probability of car, light truck and non-motorists crashes in the STAZs with higher number of shopping centers. The results indicate that the presence of more restaurants in a STAZ is positively associated with van and other motorized vehicle crashes. The results associated with parks/recreational centers show positive association with road traffic related crash risk. However, the variable effect is not significant in NB model component for other motorized vehicle crashes. Further, proportion of urban area is found to be significant in the count model component for car, light truck, van and non-motorist. In zones with larger proportion of urban areas, higher traffic related crashes are likely to occur, plausibly indicating higher density of transport activities and in turn higher traffic conflicts within an urbanized environment (see (*34*, *35*) for similar results).

Roadway Attributes. In the crash count component for van, we find that in the presence of more local roads in a STAZ, the possibility of crash risk for vans decreases. At the same time, the results associated with roadway class show that car, non-motorist, light truck, and other motorized vehicles' crash risks are positively correlated with a higher proportion of major road.

Traffic Characteristics. With respect to traffic characteristics, both AADT and truck AADT are found to have significant influence on crash occurrences for different road user groups. The model estimation results indicate that traffic related crashes are positively associated with higher AADT at the zonal level for both motorist and non-motorist road user groups. The result is in line with previous studies and can be attributable to higher exposure and/or adaptation of road users to different levels of traffic volume (see (*14*, *36*, *37*) for similar results).

Table 1. Sample Statistics for the State of Florida

Variable	Variable description	Zonal			
		Minimum	Maximum	Mean	Standard deviation
Dependent variables					
Car crashes per STAZ	Total number of car-involved crashes per STAZ	0.000	628.000	22.487	45.824
Light truck crashes per STAZ	Total number of light truck-involved crashes per STAZ	0.000	81.000	4.237	7.561
Van crashes per STAZ	Total number of van-involved crashes per STAZ	0.000	62.000	1.859	3.833
Other motorized vehicle crashes per STAZ	Total number of other motorized vehicle-involved crashes per STAZ	0.000	148.000	3.195	6.102
Non-motorist crashes per STAZ	Total number of non-motorist involved crashes per STAZ	0.000	187.000	6.557	12.830
Independent variables					
Land-use characteristics					
Shopping centers	Count of shopping centers in STAZ/10	0.000	18.900	0.583	1.049
Restaurants	Count of restaurants in STAZ/10	0.000	11.000	0.410	0.679
Park and recreational centers	Count of park and recreational centers in STAZ/10	0.000	5.400	0.079	0.167
Proportion of urban area	Total urban Area in STAZ/Total area of STAZ	0.000	1.000	0.731	0.425
Roadway attributes					
Proportion of local roads	Length of local roads in STAZ/Total length of roads in STAZ	0.000	1.000	0.085	0.192
Proportion of major roads	Length of major road in STAZ/Total length of roads in STAZ	0.000	1.000	0.544	0.371
Traffic characteristics					
AADT	Total AADT of STAZ/100,000	0.000	6.044	0.721	0.732
Truck AADT	Total truck AADT of STAZ/100,000	0.000	0.611	0.028	0.044
Socio-economic characteristics					
Proportion of industrial jobs	Total number of industrial jobs in STAZ/Total number of jobs in STAZ	0.000	1.000	0.506	0.484
Proportion of retail jobs	Total number of retail jobs in STAZ/Total number of jobs in STAZ	0.000	1.000	0.276	0.277
Proportion of households with zero vehicle	Total number of households with no vehicles in STAZ/Total number of households in STAZ	0.000	1.000	0.094	0.111
Proportion of households with one vehicle	Total number of households with one vehicle in STAZ/Total number of households in STAZ	0.000	0.967	0.411	0.135
Demographic characteristics					
Proportion of Hispanic population	Total number of Hispanic population in STAZ/Total number of population in STAZ	0.000	1.000	0.171	0.202
Proportion of Caucasian population	Total number of Caucasian population in STAZ/Total number of population in STAZ	0.000	1.000	0.621	0.278

Table 2. Multivariate Count Model Estimation Results: Gumbel Copula

	Car	Light truck	Van	Other motorized vehicle	Non-motorist
	Estimate	Estimate	Estimate	Estimate	Estimate
Variable	t-stat	t-stat	t-stat	t-stat	t-stat
Constant	1.311	0.680	–0.571	0.051	0.474
	23.940	21.571	–14.563	2.042	16.514
Land-use characteristics					
Shopping centers	0.159	0.097	–[a]	–	0.140
	8.402	6.575	–	–	9.180
Restaurants	–	–	0.080	0.127	–
	–	–	4.297	7.374	–
Park and recreational centers	0.541	0.347	0.427	–	0.599
	6.941	5.092	5.584	–	8.579
Proportion of urban area	0.570	0.074	0.562	–	0.444
	20.962	3.381	20.352	–	17.996
Roadway attributes					
Proportion of local roads	–	–	–0.298	–	–
	–	–	–4.928	–	–
Proportion of major roads	0.312	0.293	–	0.468	0.373
	10.994	10.810	–	17.745	13.489
Traffic characteristics					
AADT	0.851	0.660	0.643	0.553	0.809
	29.823	26.507	27.542	24.236	31.662
Truck AADT	–2.187	–2.225	–	2.876	–2.645
	–6.340	–7.676	–	10.052	–8.325
Socio-economic characteristics					
Proportion of industrial jobs	–0.066	0.148	–	–	–
	–3.335	7.527	–	–	–
Proportion of retail jobs	0.395	0.600	0.543	0.388	0.478
	9.146	14.451	11.727	9.891	11.001
Proportion of households with zero vehicle	–	–	–	–	0.395
	–	–	–	–	4.703
Proportion of households with one vehicle	0.752	–	0.715	0.888	–
	9.442	–	9.130	15.494	–
Demographic characteristics					
Proportion of Hispanic population	0.645	0.124	–	–	0.578
	9.434	2.971	–	–	11.405
Proportion of Caucasian population	–0.143	–	–	–	–
	–3.473	–	–	–	–
Overdispersion parameter	1.864	1.877	2.028	1.887	1.840
	76.517	63.394	48.290	58.978	64.363
Correlation parameter			3.395		
			122.559		

[a]Variable insignificant at 90% significance level.

Further, the effect of zonal level truck AADT has significant influence on all NB model components other than van. The model results reveal that the higher truck AADT at the zonal level are likely to reduce crash propensities for car, light truck and non-motorists (see (*38*) for similar result). The result may be explained by the overall cautious travel behavior of different road user groups in the presence of high volumes of heavy vehicles. On the other hand, the model estimation shows a positive correlation between truck AADT and crashes involving other motorized vehicles; perhaps indicating higher exposure of heavy vehicle and bus in these zones.

Socio-Economic Characteristics. In terms of the proportion of jobs by industry, the result associated with industrial jobs indicates that in zones with a higher proportion of industrial jobs the risk of crashes involving light trucks increases. It is likely that zones with a higher proportion of industrial jobs experience higher usage of light trucks for industrial job-related activities. On the other hand, an increase in the proportion of industrial jobs in a zone decreases the likelihood of crashes in the auto and road user groups. Zones with higher numbers of retail jobs are likely to see higher numbers of crashes involving both the motorized and non-motorized road user groups. Levine et al. (*39*) found similar impacts of retail jobs on total crash count events. Further, the risk of crashes involving non-motorists are found to be higher for STAZs with higher proportions of households without access to private vehicles. The variable is a surrogate indicator for non-motorist exposure. Household members with no private vehicles are likely to walk or cycle for daily activities resulting in higher exposure and in turn higher risk of crashes. As expected, car, van, and other motorized vehicle crash risks are found to be higher for STAZs with higher proportions of households with one available private vehicle.

Demographic Characteristics. From Table 2, we can see that the proportion of the zonal level population by ethnicity are found to be significant determinants of zonal level crash risk in count model components for car, light truck, and non-motorist crashes. Road traffic crashes for car, light truck, and non-motorists increase with an increasing proportion of Hispanic population, a result also observed in Lee et al. (*40*). On the other hand, the estimation results indicate that STAZs with greater proportions of Caucasian population are likely to experience fewer auto crashes.

Dependence Effect. As indicated earlier, the estimated Gumbel copula-based multivariate NB model provides the best fit in incorporating the correlation among different road user groups' crash count events. An examination of the copula parameter presented in the last row panel of Table 2 highlights the presence of common unobserved factors affecting zonal level crash counts of different road user groups considered in current study context. For the Gumbel copula, the dependency is entirely positive and the coefficient sign and magnitude reflects whether a variable increases or reduces the dependency and by how much. The proposed framework, by allowing for such correlation, allows us to improve data fit.

Predictive Performance Evaluation and Policy Analysis

In an effort to assess the predictive performance of the estimated models (Gumbel copula and independent models), we also perform computation of several in-sample goodness-of-fit measures. To evaluate the predictive performance of these models, we employ two different fit measures: mean prediction bias (MPB) and mean absolute deviation (MAD) both at the aggregate and disaggregate level (see (*10*) for a discussion on computing these measures). At the aggregate/disaggregate level, the computed values of MPB (MAD) for copula and independent models are 4.890 (9.071)/24.448 (45.356) and 9.529 (15.019)/47.645 (75.096), respectively. The resulting fit measures for comparing the predictive performance clearly indicate that the multivariate copula-count model offers superior predictions compared with independent count models both at the aggregate and disaggregate levels in the current study context.

The parameter effects of exogenous variables in Table 2 do not directly provide the magnitude of the effects on zonal level crash counts across different road user groups involved in crashes. For this purpose, we compute aggregate level "elasticity effects" for all the exogenous variables by using the Gumbel copula model estimates. We investigate the effect as the percentage change in the expected total zonal crash counts across different road user groups due to the change in the exogenous variable. Road user group specific elasticities would allow us to identify policy measures targeting each group separately. However, it might also be useful in identifying the contributions of exogenous variables on total crashes considering contributions from all count components. Total and group specific elasticity effects would allow us to prioritize the safety improvement programs based on the level (all groups need attention in a specific area) and type (a specific group needs attention in a specific area) of safety issues. Therefore, we also present the overall total crash elasticities in our current study. Total crash elasticities are computed by considering the change in exogenous variables across all count components simultaneously.

The computed elasticities are presented in the first row panel of Table 3 (see (*41*) for a discussion on the methodology for computing elasticities). In calculating the expected percentage change of crash counts, we increase the value of variables by 10% for each STAZ. The numbers in Table 3 may be interpreted as the percentage change in the expected total zonal crash counts due to the change in the exogenous variable. For instance, the elasticity effects for shopping centers in the car model for in-sample data indicates that, the expected mean car crashes will increase by 3.074% with an increase in 10% of shopping centers. To emphasize policy repercussions based on most critical contributory factors, we also rank each variable based on its contribution in increasing the elasticity effects – with 1 as the highest contributor and 14 as the lowest contributor across different variables considered. The results of this ranking are presented in the second row panel of Table 3.

The following observations can be made based on the results presented in Table 3. First, the most significant variable in terms of the increase in the expected number of crashes across all road user groups is AADT, which is also the most important contributor for overall road traffic crashes. Second, the ranking of variables is different across the different road user groups, which illustrates that the relative contributions

Table 3. Elasticity Effects for Multivariate Copula-Count Model for Florida

Variable	Car	Light truck	Van	Other motorized vehicle	Non-motorist	Total
			Elasticity effects			
AADT	17.371	9.483	9.986	8.467	15.055	60.362
Proportion of households with one vehicle	3.354	–	3.169	3.927	–	10.450
Shopping centers	3.074	1.193	–	–	2.400	6.668
Proportion of major roads	1.253	1.083	–	1.788	1.480	5.605
Proportion of retail jobs	0.946	1.260	1.156	0.826	1.108	5.296
Proportion of Hispanic population	1.541	0.240	–	–	1.313	3.095
Park and recreational centers	0.928	0.422	0.590	–	1.015	2.954
Restaurants	–	–	0.640	0.979	–	1.618
Proportion of urban area	0.207	0.030	0.225	–	0.171	0.634
Proportion of households with zero vehicle	–	–	–	–	0.382	0.382
Proportion of industrial jobs	–0.022	0.229	–	–	–	0.207
Proportion of local roads	–	–	–0.174	–	–	–0.174
Proportion of Caucasian population	-0.698	–	–	–	–	-0.698
Truck AADT	–1.324	–0.994	–	1.900	–1.451	–1.869

Ranking of variables based on contribution in elasticity effects

Variable	Car	Light truck	Van	Other motorized vehicle	Non-motorist	Total
AADT	1	1	1	1	1	1
Proportion of households with one vehicle	2	9	2	2	9	2
Shopping centers	3	3	7	9	2	3
Proportion of major roads	5	4	8	4	3	4
Proportion of retail jobs	6	2	3	6	5	5
Proportion of Hispanic population	4	6	9	10	4	6
Park and recreational centers	7	5	5	7	6	7
Restaurants	9	10	4	5	10	8
Proportion of urban area	8	8	6	8	8	9
Proportion of households with zero vehicle	11	12	11	12	7	10
Proportion of industrial jobs	12	7	10	11	11	11
Proportion of local roads	10	11	14	14	13	12
Proportion of Caucasian population	13	13	12	13	12	13
Truck AADT	14	14	13	3	14	14

of different exogenous variables are substantially different across different road user groups. This has important implications in identifying critical factors for crash occurrences at a zonal level. For instance, targeted policy measures should be implemented to reduce overall crashes for zones with higher AADT. However, to improve car, van, and other motorized vehicle safety, zones with higher proportions of households with one vehicle should be the major focus. On the other hand, to improve the safety of non-motorists, zones with more shopping centers should be targeted. While for improved safety of light truck traffic, zones with higher proportions of retail jobs should be the focus. Moreover, the results indicate that there are considerable differences in the elasticity effects across different road user groups, thus illustrating the value of examining separate risk factors for different road user groups. Third, the impacts, in magnitude, are substantially different in crash count events among different passenger vehicles (car, light truck, and van) for many variables. The effects are different in magnitude and a sign of the proportion of industrial jobs. These differences clearly highlight that each road user group has distinct critical risk factors underscoring the importance of examining the effect of various exogenous variables on zonal level crash count events by different road user groups. Finally, the elasticity analysis assists in providing a clear picture of attribute impacts on zonal level crash counts for different road user groups. The elasticity analysis conducted provides an illustration of how the proposed model can be applied to determine the critical factors contributing to increases in crash counts.

Conclusions

The current study contributes to safety literature both methodologically and empirically. In terms of methodology, the study generalized the bivariate copula-count model for examining multivariate crash count data by formulating and estimating a multivariate copula-count model. For examining the count components of the multivariate copula model, we employed the NB regression framework. The empirical contribution of our study was to incorporate crash counts for both motorized and non-motorized road user groups while considering different types of passenger vehicle fleet categories. Specifically, we examined crash counts for car, light truck, van, other motorized vehicle (including truck, bus, and other vehicles), and crashes involving non-motorists (pedestrians and cyclists) by employing a multivariate copula-count framework.

The proposed model was estimated using zonal level (STAZ level) road traffic crash data for the state of Florida. A host of variable groups including land-use characteristics, roadway attributes, traffic characteristics, socio-economic characteristics, and demographic characteristics were considered. The empirical analysis involved the estimation of four different multivariate copula-count models: (i) Clayton; (ii) Gumbel; (iii) Frank; and (iv) Joe copulas. The Gumbel copula model offered the most superior fit for our data. Further, the comparison between copula and the independent models confirmed the importance of accommodating dependence among crash count events of different road user groups in the macro-level analysis. Further, an in-sample validation exercise was conducted to compare the performance of the independent and copula model based on different fit measures. The resulting fit measures for comparing the predictive performance clearly indicate that multivariate copula-count model offered superior predictions compared with independent count model both at the aggregate and disaggregate levels in the current study context.

In our research, to further understand the impact of various exogenous factors, aggregate level elasticity effects were computed for all the exogenous variables by using the estimates from a multivariate copula-based count model. To emphasize policy repercussions based on most critical contributory factors, we also generated a rank for each variable based on their contributions in influencing crash frequency. The elasticity effects clearly indicated that there were considerable differences across different road user groups for the same variable, thus illustrating the value of examining separate risk factors for different road user groups. Further, the impacts were substantially different in crash count events among different passenger vehicles (car, light truck, and van). The elasticity analysis conducted provided an illustration of how the proposed model can be applied to determine the critical factors contributing to increase in crash counts.

The study is not without limitations. In modeling zonal level crash risks, we did not have access to exposure data for the different road user groups considered. It would be useful to compile zonal level exposure data for different motorized and non-motorized road user groups to enhance the model frameworks developed in our work. Further, it would be interesting to examine the count components within the copula-based multivariate approach by using Zero-inflated or Hurdle models in accommodating the preponderance of zero counts (if present) while also considering more flexible copula structures. A comparison of the proposed copula-based model with other multivariate modeling approaches will be a useful future research direction.

Acknowledgments

The authors would also like to gratefully acknowledge Florida Department of Transportation (FDOT) for providing access to Florida crash data.

Author Contributions

Study conception and design: SY, NE; data collection: SY, SUM, TN and NE; analysis and interpretation of results: SY, SUM, TN and NE; draft manuscript preparation: SY and NE. All authors reviewed the results and approved the final version of the manuscript.

References

1. Subramanian, R. Motor Vehicle Traffic Crashes as a Leading Cause of Death in the United States, 2002. *Young*, Vol. 1, 2005, p. 3.
2. Blincoe, L., A. Seay, E. Zaloshnja, T. Miller, E. Romano, S. Luchter, and R. Spicer. *The Economic Impact of Motor Vehicle Crashes, 2000*. National Highway Traffic Safety Administration, Washington, D.C., 2002.
3. Lord, D., and F. Mannering. The Statistical Analysis of Crash-Frequency Data: A Review and Assessment of Methodological Alternatives. *Transportation Research Part A: Policy and Practice*, Vol. 44, No. 5, 2010, pp. 291–305.
4. Mannering, F. L., and C. R. Bhat. Analytic Methods in Accident Research: Methodological Frontier and Future Directions. *Analytic Methods in Accident Research*, Vol. 1, 2014, pp. 1–22.
5. Yasmin, S., and N. Eluru. Latent Segmentation Based Count Models: Analysis of Bicycle Safety in Montreal and Toronto. *Accident Analysis & Prevention*, Vol. 95, 2016, pp. 157–171.
6. Mannering, F. L., V. Shankar, and C. R. Bhat. Unobserved Heterogeneity and the Statistical Analysis of Highway Accident Data. *Analytic Methods in Accident Research*, Vol. 11, 2016, pp. 1–16.

7. Chamberlain, G. Analysis of Covariance with Qualitative Data. *The Review of Economic Studies*, Vol. 47, No. 1, 1980, pp. 225–238.
8. Bhat, C. R. Quasi-Random Maximum Simulated Likelihood Estimation of the Mixed Multinomial Logit Model. *Transportation Research Part B: Methodological*, Vol. 35, No. 7, 2001, pp. 677–693.
9. Barua, S., K. El-Basyouny, and M. T. Islam. Multivariate Random Parameters Collision Count Data Models with Spatial Heterogeneity. *Analytic Methods in Accident Research*, Vol. 9, 2016, pp. 1–15.
10. Yasmin, S., and N. Eluru. A Joint Econometric Framework for Modeling Crash Counts by Severity. *Transportmetrica A: Transport Science*, Vol. 14, No. 3, 2018, pp. 230–255.
11. Anastasopoulos, P. C. Random Parameters Multivariate Tobit and Zero-Inflated Count Data Models: Addressing Unobserved and Zero-State Heterogeneity in Accident Injury-Severity Rate and Frequency Analysis. *Analytic Methods in Accident Research*, Vol. 11, 2016, pp. 17–32.
12. Nashad, T., S. Yasmin, N. Eluru, J. Lee, and M. A. Abdel-Aty. Joint Modeling of Pedestrian and Bicycle Crashes: Copula-Based Approach. *Transportation Research Record: Journal of the Transportation Research Board*, 2016. 2601: 119–127.
13. Heydari, S., L. Fu, L. F. Miranda-Moreno, and L. Jopseph. Using a Flexible Multivariate Latent Class Approach to Model Correlated Outcomes: A Joint Analysis of Pedestrian and Cyclist Injuries. *Analytic Methods in Accident Research*, Vol. 13, 2017, pp. 16–27.
14. Dong, C., D. B. Clarke, X. Yan, A. Khattak, and B. Huang. Multivariate Random-Parameters Zero-Inflated Negative Binomial Regression Model: An Application to Estimate Crash Frequencies at Intersections. *Accident Analysis & Prevention*, Vol. 70, 2014, pp. 320–329.
15. Huang, H., H. Zhou, J. Wang, F. Chang, and M. Ma. A Multivariate Spatial Model of Crash Frequency by Transportation Modes for Urban Intersections. *Analytic Methods in Accident Research*, Vol. 14, 2017, pp. 10–21.
16. Jacobsen, M. R. Fuel Economy and Safety: The Influences of Vehicle Class and Driver Behavior. *American Economic Journal: Applied Economics*, Vol. 5, No. 3, 2013, pp. 1–26.
17. Alliance, A. *Light Trucks Help Fuel 2016 Sales*. https://autoalliance.org/economy/facts-about-auto-sales/. Accessed July, 2017.
18. White, M. J. The “Arms Race” on American Roads: The Effect of Sport Utility Vehicles and Pickup Trucks on Traffic Safety. *The Journal of Law and Economics*, Vol. 47, No. 2, 2004, pp. 333–355.
19. Bhat, C. R. The Maximum Approximate Composite Marginal Likelihood (MACML) Estimation of Multinomial Probit-Based Unordered Response Choice Models. *Transportation Research Part B: Methodological*, Vol. 45, No. 7, 2011, pp. 923–939.
20. Narayanamoorthy, S., R. Paleti, and C. R. Bhat. On Accommodating Spatial Dependence in Bicycle and Pedestrian Injury Counts by Severity Level. *Transportation Research Part B: Methodological*, Vol. 55, 2013, pp. 245–264.
21. Yasmin, S., N. Eluru, A. R. Pinjari, and R. Tay. Examining Driver Injury Severity in Two Vehicle Crashes – A Copula Based Approach. *Accident Analysis & Prevention*, Vol. 66, 2014, pp. 120–135.
22. Bhat, C. R., and N. Eluru. A Copula-Based Approach to Accommodate Residential Self-Selection Effects in Travel Behavior Modeling. *Transportation Research Part B: Methodological*, Vol. 43, No. 7, 2009, pp. 749–765.
23. Wang, C., M. A. Quddus, and S. G. Ison. Predicting Accident Frequency at Their Severity Levels and Its Application in Site Ranking using a Two-Stage Mixed Multivariate Model. *Accident Analysis & Prevention*, Vol. 43, No. 6, 2011, pp. 1979–1990.
24. Ayuso, M., L. Bermúdez, and M. Santolino. Copula-Based Regression Modeling of Bivariate Severity of Temporary Disability and Permanent Motor Injuries. *Accident Analysis & Prevention*, Vol. 89, 2016, pp. 142–150.
25. Nikoloulopoulos, A. K., and D. Karlis. Regression in a Copula Model for Bivariate Count Data. *Journal of Applied Statistics*, Vol. 37, No. 9, 2010, pp. 1555–1568.
26. Cameron, A. C., T. Li, P. K. Trivedi, and D. M. Zimmer. Modelling the Differences in Counted Outcomes using Bivariate Copula Models with Application to Mismeasured Counts. *The Econometrics Journal*, Vol. 7, No. 2, 2004, pp. 566–584.
27. Nikoloulopoulos, A. K., and D. Karlis. Modeling Multivariate Count Data using Copulas. *Communications in Statistics-Simulation and Computation*, Vol. 39, No. 1, 2009, pp. 172–187.
28. Yasmin, S., N. Eluru, A. R. Pinjari, and R. Tay. Examining Driver Injury Severity in Two Vehicle Crashes–A Copula Based Approach. *Accident Analysis & Prevention*, Vol. 66, 2014, pp. 120–135.
29. Trivedi, P. K., and D. M. Zimmer. Copula Modeling: An Introduction for Practitioners. *Foundations and Trends® in Econometrics*, Vol. 1, No. 1, 2007, pp. 1–111.
30. Sklar, A. Random Variables, Joint Distribution Functions, and Copulas. *Kybernetika*, Vol. 9, No. 6, 1973, pp. 449–460.
31. Sener, I. N., N. Eluru, and C. R. Bhat. On Jointly Analyzing the Physical Activity Participation Levels of Individuals in a Family unit using a Multivariate Copula Framework. *Journal of Choice Modelling*, Vol. 3, No. 3, 2010, pp. 1–38.
32. Eluru, N., R. Paleti, R. Pendyala, and C. Bhat. Modeling Injury Severity of Multiple Occupants of Vehicles: Copula-Based Multivariate Approach. *Transportation Research Record: Journal of the Transportation Research Board*, 2010. 2165: 1–11.
33. FIRES. *Quick Stats: Crash Summary*. Florida’s Integrated Report Exchange System (FIRES). https://firesportal.com/Pages/Public/QuickStats.aspx. Accessed July, 2017.
34. Huang, H., M. Abdel-Aty, and A. Darwiche. County-Level Crash Risk Analysis in Florida: Bayesian Spatial Modeling. *Transportation Research Record: Journal of the Transportation Research Board*, 2010. 2148: 27–37.
35. Siddiqui, C., M. Abdel-Aty, and K. Choi. Macroscopic Spatial Analysis of Pedestrian and Bicycle Crashes. *Accident Analysis & Prevention*, Vol. 45, 2012, pp. 382–391.
36. Milton, J. C., V. N. Shankar, and F. L. Mannering. Highway Accident Severities and the Mixed Logit Model: An

Exploratory Empirical Analysis. *Accident Analysis & Prevention*, Vol. 40, No. 1, 2008, pp. 260–266.

37. Hadayeghi, A., A. Shalaby, and B. Persaud. Macrolevel Accident Prediction Models for Evaluating Safety of Urban Transportation Systems. *Transportation Research Record: Journal of the Transportation Research Board*, 2003. 1840: 87–95.
38. Cai, Q., M. Abdel-Aty, J. Lee, and N. Eluru. Comparative Analysis of Zonal Systems for Macro-Level Crash Modeling. *Journal of Safety Research*, Vol. 61, Suppl. C, 2017, pp. 157–166.
39. Levine, N., K. E. Kim, and L. H. Nitz. Spatial Analysis of Honolulu Motor Vehicle Crashes: II. Zonal Generators. *Accident Analysis & Prevention*, Vol. 27, No. 5, 1995, pp. 675–685.
40. Lee, J., M. Abdel-Aty, and X. Jiang. Development of Zone System for Macro-Level Traffic Safety Analysis. *Journal of Transport Geography*, Vol. 38, 2014, pp. 13–21.
41. Eluru, N., and C. R. Bhat. A Joint Econometric Analysis of Seat Belt Use and Crash-Related Injury Severity. *Accident Analysis & Prevention*, Vol. 39, No. 5, 2007, pp. 1037–1049.

The Standing Committee on Statistical Methods (ABJ80) peer-reviewed this paper (18-01587).

Article

TRR
JOURNAL OF THE TRANSPORTATION RESEARCH BOARD

Transportation Research Record
2018, Vol. 2672(30) 76–89
© National Academy of Sciences:
Transportation Research Board 2018
Article reuse guidelines:
sagepub.com/journals-permissions
DOI: 10.1177/0361198118801355
journals.sagepub.com/home/trr
SAGE

Joint Modeling of Traffic Incident Duration Components (Reporting, Response, and Clearance Time): A Copula-Based Approach

Haluk Laman[1], Shamsunnahar Yasmin[1], and Naveen Eluru[1]

Abstract

The current study develops a trivariate framework that accommodates for inherent dependencies across the various components of incident duration. A unique ordered response model structure (sometimes referred to as grouped ordered response model) is introduced for modeling three durations—reporting time, response time, and clearance time—in this study. Further, as opposed to employing a simulation oriented multivariate model approach, the study proposes and estimates a copula-based methodology that allows for a closed-form probability computation. The approach is the first application of this model for incident duration analysis. The proposed copula framework is estimated to identify factors affecting incident duration components from a host of characteristics including incident characteristics, traffic conditions, roadway, and environmental characteristics. The data for the analysis are obtained from the 2015 Florida Department of Transportation incident database. The model estimates were also augmented by conducting policy analysis by generating a three-dimensional representation of incident frequencies as a function of reporting, response, and clearance time.

Traffic congestion can generally be attributed to either recurring or nonrecurring events. Congestion arising from recurring events is generally a result of mismatched transportation demand and supply (or capacity). Nonrecurring congestion, on the other hand, is a result of unexpected (or irregular) events such as abandoned vehicles, adverse weather, spilled loads, highway debris, and traffic crashes. The potential solutions for congestion arising from these two sources are vastly different. This research focuses on nonrecurrent congestion. Nonrecurring congestion is responsible for approximately one-quarter of all traffic delay on U.S. roadways (*1*). According to Roper (*2*), every minute that a freeway travel lane is blocked results in 4–5 minutes of traffic delay after the incident is cleared. The U.S. Department of Transportation Strategic Plan for Fiscal-Year 2010–2015 reports that 2.8 billion gallons of gasoline are consumed every year in the U.S. as a result of incident-related congestion events (*3*). Moreover, longer incident durations can increase the risk of secondary incidents (*4*). Consequently, transportation agencies are developing traffic incident management strategies to reduce the overall duration of incidents to minimize their impacts on travelers and the environment.

The overall incident duration, as identified by the Highway Capacity Manual (*5*), is composed of the following four phases: notification time, response time, clearance time and traffic recovery time (*6*). Incident clearance (third phase) is usually the longest component of the incident duration time (*4*). The traffic recovery time (fourth phase) is a function of total duration of the first three phases and the traffic demand on the facility. Given the importance of different phases of incident duration, the objective of the proposed research effort is to study the factors influencing incident reporting, response, and clearance times—with a goal of reducing the congestion impacts of nonrecurring events while providing improved traffic incident management plans. An important factor affecting incident clearance is the response personnel involved. In the state of Florida, in addition to the traditional agencies, a road ranger service

[1]Department of Civil, Environmental & Construction Engineering, University of Central Florida, Orlando, FL

Corresponding Author:
Address correspondence to Shamsunnahar Yasmin:
shamsunnahar.yasmin@ucf.edu

patrol assists in the incident clearance process. In particular, Florida Department of Transportation (FDOT) offers a unique service via road ranger service patrol to offer free assistance to road users on highways to reduce delay while enhancing safety for the public. Since its inception, the Road Rangers (RR) service has offered nearly 4.3 million assists (*7*). The objectives of the program include reducing traffic crashes, assisting the Florida Highway Patrol to reduce incident duration, providing assistance to disabled or stranded vehicles, removing road debris, and increasing safety at incident sites. Toward meeting these goals, the RR trucks monitor congested areas and high-incident locations of the urban expressway for different nonrecurrent incidents. This research examines incident clearance duration with a specific emphasis on the impact of RR service patrol program. The analysis would allow recommendations to be made on the performance of the RR patrol while offering recommendations for similar programs in other states.

Earlier Research

Several research efforts have examined incident duration as a function of incident characteristics, traffic conditions, and roadway characteristics. Literature in incident management focusing on incident duration is guided by two objectives. The first objective is an emphasis on incident duration prediction. The second objective is to identify the exogenous variables that affect incident duration (*8*). A detailed review of these earlier research efforts is summarized in Table 1. The information presented in the table includes the study, study region and data source, outcome variable, type of model, identifiers for different times considered (reporting time, response time), and important factors identified (classified as incident characteristics, traffic conditions, roadway, environmental characteristics, and others).

Current Study in Context

It is evident from Table 1 that previous research has provided insights on total incident duration, clearance time, and response time. However, limitations still exist in earlier work. The examination of all components of incident duration as dependent variables is critical to understanding the overall incident clearance process. The consideration of components such as reporting time as explanatory variables in modeling other duration components could potentially lead to endogeneity bias. The factors that increase reporting time (such as an accident in a remote location) could potentially lead to increased response time and clearance time. Thus, the model development process for incident duration should examine the impact of various components of incident duration within a multivariate framework. The current study develops a trivariate framework that accommodates for inherent dependencies across the various components of incident duration. Second, a unique ordered response model structure (sometimes referred to as grouped ordered response model) is introduced for modeling the three durations in this study. In the presence of large groupings of dependent variables around a particular value, a linear regression model (or a log-linear model) would perform poorly. The same drawback will apply to hazard duration models. On the other hand, the approach allowing the grouped alternatives reduces the sensitivity of the model to large groupings around a single value. The proposed grouped ordered response approach is the first application of this model for incident duration analysis.

Finally, as opposed to employing a simulation oriented multivariate model approach, a copula-based methodology that allows for a closed-form probability computation is proposed and estimated. In recent years, several studies have highlighted the values of copula models in transportation (*38–41*). The proposed copula framework is estimated to identify factors affecting incident duration components from a host of characteristics including incident characteristics, traffic conditions, and roadway and environmental characteristics. The data for the analysis are obtained from the 2015 FDOT incident database for events involving RR professionals. A model illustration exercise is conducted to highlight potential applications of the proposed model using 3-dimensional surface plots. A validation exercise is also conducted to test the predictive performance of the model.

Econometric Methodology

Copula-Based Approach

A copula is a device or function that generates a stochastic dependence relationship (i.e., a multivariate distribution) among random variables with prespecified marginal distributions (*38*, *42*). Let C be an I-dimensional copula of uniformly distributed random variables $U_1, U_2, U_3, \ldots, U_I$ with support contained in $[0, 1]^I$. Thus, the joint distribution of these random terms can be defined as.

$$C_\theta(u_1, u_2, u_3, \ldots, u_I) = Pr(U_1 < u_1, U_2 < u_2, U_3 < u_3, \ldots, U_I < u_I) \quad (1)$$

where θ is a parameter vector of the copula commonly referred to as the dependence parameter vector. Let us consider $I(\varepsilon_1, \varepsilon_2, \varepsilon_3, \ldots, \varepsilon_I)$ random variables each with univariate continuous marginal distribution function $F(Z_i) = \Pr(\varepsilon_i < Z_i)$. Thus, a joint I-dimensional

Table 1. Summary of Literature Review

Study	Study region and data source	Outcome variable	Type of model	Reporting time considered?	Response Time considered?	Important factors identified			
						Incident	Traffic	Roadway	Environmental
Golob et al., (*9*)	Los Angeles / 1983-1985	Total incident duration	Lognormal Distribution	No	No	Yes	No	No	No
Giuliano, (*10*)	Los Angeles / 1987	Freq. and total incident duration	ANOVA	No	No	Yes	Yes	No	No
Nam and Mannering, (*11*)	Washington / 1994 and 1995	Reporting, response, and clearance time	HBM: Weibull and Log-Logistic Survival	Yes	Yes	Yes	Yes	Yes	Yes
Chimba and Horne, (*12*)	Tennessee / 2004 - 2010	Total incident duration	HBM: Log-Logistic Survival Model	No	No	Yes	Yes	Yes	Yes
Tavassoli Hojati et al., (*13*)	Queensland, Australia / 2009	Log of incident duration in minutes	HBM: Log-Logistic, Lognormal, Weibull Survival	No	No	Yes	Yes	Yes	Yes
Garib et al., (*14*)	Oakland, California / 1993	Delay in vehicle hours and log of incident duration	Multiple Regression Models	No	No	Yes	Yes	Yes	Yes
Ding et al., (*15*)	Washington / 2009	Response time and clearance time	Binary probit and Switching regression models	No	Yes	Yes	Yes	No	Yes
Weng et al., (*16*)	Maryland I-95 / 2010 and 2011	Response time and clearance time	Cluster Based Lognormal Distribution Model	No	No	Yes	Yes	Yes	Yes
Khattak et al., (*17*)	Chicago / 1989-1990	Total incident duration	Truncated regression, Time sequential models	No	Yes	Yes	Yes	No	Yes
Valenti et al., (*18*)	Italy / 2005 (3 Months)	Total incident duration	L. Regression, Decision Tree, ANN, Support Vector Machine (SVM), kNN	No	No	Yes	Yes	Yes	No
Smith and Smith, (*19*)	Virginia, I-64, I-264 / 1997-2000	Clearance time	Non-parametric Regression and CART,	Yes	Yes	Yes	No	No	Yes
Demiroluk and Ozbay, (*20*)	New Jersey / 2005	Total incident duration	Three types of Bayesian Networks: Naïve Bayes, TAN, and K2	No	No	Yes	Yes	Yes	Yes
Chung, (*21*)	S. Korea / 2006 and 2007	Total incident duration	Hazard Based Duration Models (HBM): Log-Logistic AFT	No	No	Yes	Yes	Yes	No
Sullivan, (22)	Chicago, Charlotte, Houston, Orlando, SF, LA/ 1994	Total incident duration	Lognormal Distribution and IMPACT model	Yes	Yes	Yes	Yes	Yes	No

(continued)

Table 1. (continued)

Study	Study region and data source	Outcome variable	Type of model	Reporting time considered?	Response Time considered?	Important factors identified			
						Incident	Traffic	Roadway	Environmental
Hu et al., (*23*)	London / 2007-2008	Total incident duration	HBM: Log-logistic AFT Survival Model	No	No	Yes	Yes	No	Yes
Wang et al., (*24*)	United Kingdom / 2000-2001	Total incident duration	Fuzzy Logic and ANN	No	No	Yes	Yes	Yes	No
Ozbay and Noyan, (*25*)	N. Virginia / 1997	Clearance time	Bayesian Networks	No	No	Yes	No	Yes	No
Lee and Wei, (*26*)	Taiwan / 2004-2005	Response time and clearance time	Genetic Algorithm and ANN models	No	No	Yes	Yes	Yes	Yes
Wu et al., (*27*)	Netherlands, 2005	Total incident duration	Support Vector Regression (SVR)	No	No	Yes	Yes	No	Yes
Zhan et al., (*28*)	F. Lauderdale / 2006-2007	Clearance time	M5P Tree Algorithm	No	Yes	Yes	Yes	Yes	Yes
Khattak et al., (*29*)	Hampton / 2006	Total incident duration	OLS and truncated regression	No	No	Yes	Yes	Yes	Yes
Zhang and Khattak, (*30*)	Hampton Roads / 2005	Contained and extended event durations	OLS Regression	No	No	Yes	Yes	Yes	Yes
Ghosh et al., (*4*)	Michigan / 2009	Clearance time	HBM: Exp. Weibull, Lognormal, Log-logistic, Gamma	No	No	Yes	Yes	Yes	Yes
Chung and Yoon, (*31*)	California / 2001	Total incident duration	HBM: Lognormal AFT	No	No	Yes	Yes	Yes	Yes
Al Kaabi et al., (*32*)	Abu Dhabi, UAE / 2009	Response time	HBM: Weibull, Lognormal, Log-Logistic AFT	No	Yes	Yes	Yes	Yes	Yes
Vlahogianni and Karlaftis, (*33*)	Athens, Greece / 2012	Total incident duration	Partial Logistic Regression ANN	No	No	Yes	Yes	Yes	Yes
Pereira et al., (*8*)	Singapore / 2010-2011	Total incident duration	LDA Topic Modelling / LR, SVM, ANN, DT	No	No	Yes	Yes	Yes	Yes
Tavassoli Hojati et al., (*34*)	Queensland 2011	Incident duration, recovery time	HBM: Weibull, Log-logistic AFT	No	No	Yes	Yes	Yes	Yes
Li, (*35*)	Beijing, China / 2008	Preparation, travel, clearance time	HBM: Gamma, Weibull AFT	No	Yes	Yes	Yes	No	Yes
Zou et al., (*36*)	Washington I-5 / 2009	Clearance time	HBM: Weibull, Log-logistic, Lognormal AFT	No	Yes	Yes	Yes	Yes	Yes
Ma et al., (*6*)	Washington I-5 / 2012	Clearance time	Gradient Boosting Decision Tree	No	Yes	Yes	Yes	Yes	Yes
Zhu et al., (*37*)	Washington / 2011-2013	Incident duration	kNN / HBM: Log-logistic, lognormal, Weibull AFT	No	No	Yes	Yes	No	No

distribution function of the random variables with the continuous marginal distribution functions $F(Z_i)$ can be formed as follows (*43*):

$$F(Z_1, Z_2, Z_3, \ldots, Z_I) = \Pr(\varepsilon_1 < Z_1, \varepsilon_2 < Z_2, \varepsilon_3 < Z_3, \ldots, \varepsilon_I < Z_I) = C_\theta[u_1 = F(Z_1), u_2 = F(Z_2), u_3 = F(Z_3), \ldots, u_I = F(Z_1)) \quad (2)$$

The specification defined in Equation 2 offers an approach to develop different dependency patterns for the random variables $(\varepsilon_1, \varepsilon_2, \varepsilon_3, \ldots, \varepsilon_I)$ based on the copula that is used as the underlying basis of construction.

Ordered Group Response Framework

Let q be an index for events (incidents in the current empirical context) $(q = 1, 2, \ldots, Q)$, and let j be the index for duration components $(j = 1, 2, \ldots, J)$. Also, let k_j be an index for the time intervals $(k_j = 1, 2, \ldots, K_j)$ for the duration components. In this study, j takes the values of reporting time $(j = 1)$, response time $(j = 2)$ and clearance time $(j = 3)$. In the usual ordered response framework notation, one can write the latent propensity (y^*_{qj}) of duration component j for incident q to take a time interval level as a function of relevant covariates, and then relate this latent propensity to the time interval (y_{qj}) representing the time interval elapsed for duration component j in the event of incident q through threshold bounds as:

$$y^*_{qj} = \boldsymbol{\beta}_j \boldsymbol{x}_{qj} + \boldsymbol{\sigma}_{k_j} + \varepsilon_{qj}, y_{qj} = k_j, if\, \psi_{k_j} < y^*_{qj} < \psi_{k_j+1} \quad (3)$$

where $\boldsymbol{x}_{qj}$ is a vector of exogenous variables for duration component j in incident q, $\boldsymbol{\beta}_j$ is a corresponding vector of coefficients to be estimated, and ψ_{k_j} is the lower bound threshold for grouped time interval level k_j specific to duration component j. The ε_{qj} term captures the idiosyncratic effect of all omitted variables for duration component j for the event q. Further, $\boldsymbol{\sigma}_{k_j}$ is a vector of time interval category-specific coefficients for time interval alternative k_j in duration component j. The ε_{qj} terms are assumed identical across duration components, each with a univariate continuous marginal distribution function $F(Z_{qj}) = \Pr(\varepsilon_{qj} < Z_{qj})$. The error terms are assumed to be independently logistic distributed with variance $\boldsymbol{\lambda}^2_{qj}$. The variance vector is parameterized as follows:

$$\boldsymbol{\lambda}_{qj} = \exp(\delta + \varrho \boldsymbol{w_{qj}}) \quad (4)$$

where δ is a constant, $\boldsymbol{w_{qj}}$ is a set of exogenous variables associated with duration component j in incident q and ϱ is the corresponding vector of parameters to be estimated. The parameterization allows for the variance to be different across events accommodating for heteroscedasticity. Given these relationships across the different parameters, the probability for duration component j for time interval in category k_j is given by:

$$\Pr(y_{qj} = k_j) = \phi\left(\frac{\psi_{k_j+1} - (\boldsymbol{\beta}_j \boldsymbol{x}_{qj} + \boldsymbol{\sigma}_{k_j})}{\boldsymbol{\lambda}_{qj}}\right) - \phi\left(\frac{\psi_{k_j} - (\boldsymbol{\beta}_j x_{qj} + \boldsymbol{\sigma}_{k_j})}{\boldsymbol{\lambda}_{qj}}\right) \quad (5)$$

where $\phi(\cdot)$ is the standard logistic distribution function.

Joint Model Formulation and Estimation

In examining the grouped time intervals across different duration components simultaneously, the levels of correlations between three dimensions of interests depend on the type and extent of dependency among the stochastic terms (ε_{qj}) of Equation 3. Thus, dependence in the ε_{qj} terms across duration components j in the same event q is accommodated to allow for the unobserved cluster effects and the copula-based joint probability function as presented in Equation 2 can be expressed as:

$$F(Z_{q1}, Z_{q2}, Z_{q3}) = \Pr(\varepsilon_{q1} < Z_{q1}, \varepsilon_{q2} < Z_{q2}, \varepsilon_{q3} < Z_{q3}) = C_{\theta_q}[u_{q1} = F(Z_{q1}), u_{q2} = F(Z_{q2}), u_{q3} = F(Z_{q3})) \quad (6)$$

where C_{θ_q} is the copula density. It is important to note here that the level of dependence between grouped time interval levels across different duration components can vary across clusters. Therefore, in the current study, the dependence parameter θ_q is parameterized as a function of observed attributes as follows:

$$\theta_q = fn(\boldsymbol{\gamma s}_q) \quad (7)$$

where $\boldsymbol{s}_q$ is a column vector of exogenous variables, $\boldsymbol{\gamma}$ is a vector of unknown parameters (including a constant) and fn represents the functional form of parameterization. Based on the dependency parameter permissible ranges, alternate parameterization forms for the four copulas are considered in this analysis. For the Clayton and Frank copulas the study employs $\theta_q = \exp(\boldsymbol{\gamma s}_q)$, and for Joe and Gumbel copulas the study employs $\theta_q = 1 + \exp(\boldsymbol{\gamma s}_q)$ (see other work (*44–46*) for a similar approach).

Thus, the probability of the observed vector of grouped time interval levels across different duration components in incident q can be written as:

$$\Pr(y_{q1} = k_{q1}, y_{q2} = k_{q2}, y_{q3} = k_{q3}) = \int_{K_q} C_{\theta_q}\left(F(y^*_{q1}), F(y^*_{q2}), F(y^*_{q3})\right) dy^*_{q1} dy^*_{q2} dy^*_{q3} \quad (8)$$

where the integration domain K_q is simply the multivariate region of the y^*_{qj} variables determined by the observed vector of grouped time interval variables (k_{q1}, k_{q2}, k_{q3}). The dimensionality of the integration, in general, is equal to the number of duration components in the incident. In the current study context, event-level clusters with identical dependencies between pairs of duration components in the incident are considered. The cluster size in this study is 3, and allows estimation of the formulation presented in Equation 6 in a closed-form structure (see Eluru et al. [*47*] for detailed discussion of computation issue with clusters greater than 5). The probability in Equation 6 can be written in terms of 2^3 closed-form multivariate cumulative distribution functions as follows:

$$
\begin{aligned}
&\Pr(y_{q1} = k_{q1}, y_{q2} = k_{q2}, y_{q3} = k_{q3}) \\
&= Pr(\psi_{k_{q1}} < y^*_{q1} < \psi_{k_{q1}+1}, \psi_{k_{q2}} < y^*_{q2} < \psi_{k_{q2}+1}, \psi_{k_{q1}} < y^*_{q3} < \psi_{k_{q3}+1}) \\
&= \sum_{a_1=1}^{2} \sum_{a_2=1}^{2} \sum_{a_3=1}^{2} (-1)^{a_1+a_2+a_3} \\
&\left[Pr(y^*_{q1} < \psi_{k_{q1}+a_1-1}, y^*_{q2} < \psi_{k_{q2}+a_2-1}, y^*_{q3} < \psi_{k_{q3}+a_3-1})\right] \\
&= \sum_{a_1=1}^{2} \sum_{a_2=1}^{2} \sum_{a_3=1}^{2} (-1)^{a_1+a_2+a_3} \\
&\left[C_{\theta_q}\left(u_{k_{q1}+a_1-1}, u_{k_{q2}+a_2-1}, u_{k_{q3}+a_3-1}\right)\right]
\end{aligned}
\tag{9}
$$

The parameters to be estimated in the model may be gathered in a vector $\Omega = \left(\boldsymbol{\beta}_j, \boldsymbol{\sigma}_{k_j}, \psi_{k_j}, \delta, \varrho, \gamma\right)$. The likelihood function for incident q may be constructed based on the probability expression in Equation 9 as:

$$L_q(\Omega) = \Pr(y_{q1} = k_{q1}, y_{q2} = k_{q2}, y_{q3} = k_{q3}) \tag{10}$$

The likelihood function to be maximized is then given by:

$$L(\Omega) = \prod_q L_q(\Omega) \tag{11}$$

This analysis employs four Archimedean copulas, Frank, Clayton, Joe and Gumbel copulas (a detailed discussion of these copulas is available in Bhat and Eluru [*38*]). The estimation routines for the models in current study are coded in GAUSS Matrix Programming software (*48*).

Data Description

This paper focuses on Central Florida traffic incidents provided by FDOT that occurred in year 2015 with participation from at least one RR patrol. In 2015; RR responded to a total of 57,238 incidents. Through an extensive data preparation process, only those events in which complete information was available for reporting time, response time, and clearance time were selected. Subsequently, the final compiled dataset consisted of 50,319 events. From the final dataset, 14,870 records are randomly sampled for the purpose of model estimation.

The first row panel of Table 2 presents the dependent variables statistics considered in the current study context. The study considers five, eight, and eleven categories of reporting, response and clearance time, respectively. With respect to independent variables, the attributes considered can be grouped into four broad categories: incident characteristics, traffic characteristics, roadway characteristics and environmental characteristics. Descriptive statistics of variables considered in this modeling exercise are provided in the second row panel of Table 2.

Empirical Analysis

Optimal Copula Model Selection

The empirical analysis involves four different families of copula models estimation to explain the dependence between reporting, response, and clearance times of traffic incidents. An independent copula model (separate ordered group response models for incident durations) is estimated to establish a benchmark for copula structure selection. The copula models estimated in this analysis include: (1) Clayton, (2) Gumbel, (3) Frank, and (4) Joe copulas. For each of the selected copula structures, models with and without parameterization in the dependence effects were also estimated (see Equation 7). Thus, a comparison of 9 models was undertaken. In order to determine the optimal copula model, performances of the alternative copula models were tested by log-likelihood values as well as employing the Bayesian information criteria (BIC). The copula with the lowest BIC is the preferred model. The BIC values for the various estimated models are presented in Table 3. Please note that the Clayton copula models collapsed to the independent model (thus instead of 9 rows, there are only 7 rows). Based on the model fit comparison, several copula models offer improved fit relative to the independent model, supporting the hypothesis that those three components of incident duration exhibit strong dependency. Among the copula models, the Gumbel-parameterized model offers the most significant improvement in log-likelihood and lower BIC. Henceforth, the Gumbel-parameterized model structure is described in detail.

Estimation Results

In this section, the results for the Gumbel-parameterized copula-based model are presented. In Table 4, the first, second, and third column panels of results correspond to

Table 2. Sample Statistics

Variable	Definition	Statistics		
Dependent variables				
Attributes		Min.	Max.	Average
System level characteristics				
Reporting time	*Notified Time – Detected Time (minutes)*	0.00	244.21	0.13
Response time	*Arrival Time – Notified Time (minutes)*	0.00	392.93	1.79
Clearance time	*Closed Time – Arrival Time (minutes)*	0.00	8924.00	81.81
Group-specific characteristics				
Attributes			Frequency	
Reporting time				
T1-Bin 1	>0~0.5 minutes		14551	
T1-Bin 2	>0.5~1 minutes		134	
T1-Bin 3	>1~1.5 minutes		52	
T1-Bin 4	>1.5~2 minutes		22	
T1-Bin 5	>2 minutes		111	
Response time				
T2-Bin1	>0~5 minutes		13828	
T2-Bin2	>5~10 minutes		201	
T2-Bin3	>10~15 minutes		206	
T2-Bin4	>15~20 minutes		153	
T2-Bin5	>20~30 minutes		200	
T2-Bin6	>30~40 minutes		127	
T2-Bin7	>40~50 minutes		57	
T2-Bin8	>50 minutes		98	
Clearance time				
T3-Bin1	0~5 minutes		6469	
T3-Bin2	>5~10 minutes		2096	
T3-Bin3	>10~15 minutes		1334	
T3-Bin4	>15~20 minutes		961	
T3-Bin5	>20~40 minutes		1526	
T3-Bin6	>40~60 minutes		592	
T3-Bin7	>60~80 minutes		291	
T3-Bin8	>80~100 minutes		226	
T3-Bin9	>80~120 minutes		140	
T3-Bin10	>120~140 minutes		98	
T3-Bin11	>140 minutes		1137	
Independent variables				
Continuous predictors	Minimum/Maximum		Average	
Distance from CBD	0.003/0.278		0.078	
Categorical predictors	Frequency		Percentage (%)	
Incident characteristics / Activity type				
Number of vehicles involved				
More than one vehicle	487		3.3	
One vehicle	14,383		96.7	
Event type				
Crash	1,102		7.4	
Disabled vehicle	8,683		58.4	
Other event	5,085		34.2	
Activity type				
Tire service	2,304		15.5	
Abandoned activity	1,348		9.1	
Debris activity	3,498		23.5	
Mechanical activity	1,997		13.4	
Other activity	5,723		38.5	
Notifier agency				
Notified by police	828		5.6	
Notified by road ranger	1,254		8.4	
Other agency	12,788		86	
RR agency responded				
I-4 road ranger responded	6,315		42.5	

(continued)

Table 2. (continued)

Variable	Definition	Statistics
CFX road ranger responded	7,908	53.2
Other RR agency	647	4.3
Traffic characteristics		
Time of the day		
Morning peak	3,782	25.4
Evening peak	4,432	29.8
Nighttime	1,720	11.6
Day of the week		
Weekday	10,725	72.1
Weekend	4,145	27.9
Roadway characteristics		
Event location		
Intersection	4,922	33.1
Non-intersection	9,948	66.9
Built environment		
Rural	1,511	10.2
Urban	13,359	89.8
County		
Seminole	1,269	8.5
Other counties	13,601	91.5
Roadways		
At Interstate-4	6,691	45
At State Road 417	1,682	11.3
At State Road 429	1,210	8.1
At State Road 408	3,338	22.4
Other roadways	1,949	13.2
Environmental characteristics		
Months		
July	1,379	9.3
August/September	1,269	8.5
September	1,197	8
Other months	12,222	82.2

Table 3. Log-Likelihood and BIC of Copula Models

Models	Number of parameters	Log-likelihood	BIC
Gumbel-parameterized	55	−33045.90	*66620.19*
Joe-parameterized	55	−33060.92	66650.23
Frank-parameterized	55	−33081.59	66691.56
Gumbel	53	−33108.20	66725.58
Frank	53	−33111.92	66733.02
Joe	53	−33129.62	66768.41
Independent	56	−33152.22	66842.44

the reporting, response, and clearance time. For the ease of model discussion, the impact of an explanatory variable is presented for all three components of the joint model at the same time. A positive (negative) coefficient associated with a variable indicates propensity for longer (shorter) duration.

Incident Characteristics. The results highlight a variation in duration based on event type. Specifically, major incidents such as incidents involving multiple vehicle or crash events tend to have longer response times (*11*). On the other hand, minor incidents such as disabled vehicles have lower response and clearance times. Abandoned vehicle and debris events are found to be associated with shorter response delays compared with other activities. Tire services such as a "tire change" lead to longer response time. Mechanical activity-associated events (such as engine, gas, overheating activities) are found to be significantly associated with shorter clearance time.

In terms of notification personnel, if the incident is notified by a police officer, reporting time reduces significantly. According to the scale component results; when the notifier is a RR, incident clearance time increases. RR patrols are categorized as I-4, Broward, Palm Beach, and Central Florida Expressway Authority (CFX) RR. Among these categories, I-4 RR and CFX RR are associated with longer response delays compared with other RR. It is possible that the presence of heavy traffic on these facilities increases response times.

Traffic Characteristics. With respect to traffic characteristics, model results show a significant reduction in reporting time of incidents occurring during morning peak hours (see Nam and Mannering [*11*] for a similar result). On the other hand, response time tends to increase in evening rush hours. The results in the literature are ambiguous in this regard, with supporting as well as contradictory evidence (*11*, *49*). The increase in response time can be attributed to increased travel delay resulting in longer travel times for responders. Clearance time is likely to increase in both morning and evening peak hours. Previous research is consistent with this finding (*11*). Response and clearance times of nighttime (10:00 p.m. to 6:00 a.m.) incidents tend to take longer compared with response and clearance time of daytime (6:00 a.m. to 10:00 p.m.) incidents. Traffic incidents occurring on weekends have a tendency to reduce incident clearance time as compared with weekday incidents, which is consistent with the findings from earlier literature (*19*, *50*).

Roadway Characteristics. It can be observed from the results that location of crashes has significant effects on the incident duration phases. Reporting and response time for events that occur within Seminole County was found to take longer than other counties. Based on data from FDOT crash statistics, Seminole County is one of the safest counties in the state (*51*). Thus, the resource allocation of the county toward incident management might be lower than other counties. Furthermore, the responder agencies in the vicinity of Seminole county are I-4 RR and CFX RR. Given that these major roadways are further from Seminole county relative to other counties could potentially increase reporting and response times. Moreover, among the highways analyzed in this model, incidents that occurred on Interstate highway-4 (I-4) tend to have longer response and clearance times. The significant traffic volume on the roadway is a plausible reason for these findings.

Central Florida region is home to several tollways. Among these, tollway state road (SR) 417 incidents are found to be associated with longer clearance time, whereas toll road SR 429 incidents show significant association with shorter clearance time. The results are intuitive given the traffic volumes observed on these facilities. SR-417 is a 55 mile stretch of tollway with about 375,000 vehicles served per day. SR-429 is a 23 mile stretch of tollway with about 123,000 vehicles served per day. Thus, responding personnel have easier access for SR 429 reducing clearance time. Traffic events at intersections take longer reporting, response, and clearance times. The distance from incident location to the Central Business District (CBD) area was considered in the model. According to the results, longer distances for CBD tended to have longer response delays and clearance durations. This finding is consistent with a study effort which indicates that longer distances from CBD tend to increase the total incident duration (*13*).

Environmental Characteristics. Among the months considered in the models, summer months were found to significantly influence the reporting and response times. For example, when incidents occur in July, August, or September, reporting and response times reduce relative to reporting and response times for all other months. The reasons for the differential impacts in these months is not immediately apparent. It is possible that during these months, traffic volumes in Central Florida region are lower as the universities and schools are closed. The result warrants future investigation.

Alternative Specific Effects. In the grouped ordered specification of the joint model, alternative specific constants for categories considered across different duration components were also estimated. It is worthwhile to mention here that it is possible to estimate group-specific effects for each group considered across different duration components. However, in the joint model specifications, group-specific effects were estimated if they improve data fit. The results of these group-specific effects are presented in the second row panel of Table 4. For the reporting time component, none of the group-specific effects improved data fit further and thus are not included in the final model specifications. With respect to response and clearance time, group-specific components are estimated for one (T2-Bin2) and four (T3-Bin1, T3-Bin2, T3-Bin3 and T3-Bin4) categories, respectively. Adding more group-specific components did not improve the data fit further in the current study context, and therefore are not included in the final joint model specifications. These parameters are similar to constants in discrete choice models and do not really have a substantive interpretation.

Variance Component. As indicated earlier, the variance of the grouped ordered components are estimated as a

Table 4. Gumbel-Parameterized Copula Model Results

Gumbel-parameterized copula ordered logit—Generalized ordered logit model

	Reporting time		Response time		Clearance time	
Variable	Estimate	t-stat	Estimate	t-stat	Estimate	t-stat
Constant	−3.975	−13.28	−37.502	−15.18	−141.212	−16.31
Incident characteristics / Activity type						
Number of vehicles involved (Base: One vehicle)						
More than one vehicle	0.828	4.56	-	-	-	-
Event type (Base: Other event)						
Crash	2.146	11.13	-	-	-	-
Disabled vehicle	-	-	−46.768	−14.32	−36.978	−10.55
Activity type (Base: Other activity)						
Tire service	-	-	5.96	2.62	-	-
Abandoned activity	-	-	−45.542	−9.34	-	-
Debris activity	-	-	−13.885	−7.78	-	-
Mechanical activity	-	-	-	-	−34.652	−6.73
Notifier agency (Base: Other agency)						
Notified by police	−13.671	−2.1	-	-	-	-
Notified by road ranger	-	-	-	-	−17.161	−3.09
Traffic characteristics						
Time of the day (Base: Other time)						
Morning peak	−0.477	−2.37	-	-	21.035	5.16
Evening peak	-	-	9.134	6.57	19.787	4.11
Nighttime	-	-	7.397	4.43	48.511	9.3
Day of the week (Base: Weekdays)						
Weekend	-	-	-	-	−8.677	−2.7
Roadway characteristics						
Distance from CBD	-	-	0.005	3.61	0.015	5.17
Event location (Base: Non-intersection)						
Intersection	0.509	3.2	6.519	3.24	12.135	3.3
County (Base: Other counties)						
Seminole	0.67	3.28	5.397	2.97	-	-
Roadways (Base: Other roadways and State 408)						
At Interstate-4	-	-	22.694	13.37	52.956	12.24
At State Road 417	-	-	-	-	32.706	6.31
At State Road 429	-	-	-	-	−39.833	−6.42
Environmental characteristics						
Months (Base: Other months)						
July	−0.485	−1.65	−6.366	−2.86	-	-
August/September	−0.473	−2.28	-	-	-	-
September	-	-	−6.898	−2.91	-	-
Variance components						
Constant	−0.273	−2.74	2.771	69.24	4.248	111.11
At intersection	-	-	−0.092	−2.12	-	-
At rural area (Base: Urban)	-	-	-	-	−0.106	−3.51
At State Road 408	-	-	-	-	0.182	6.88
Disabled vehicle	-	-	0.414	7.64	-	-
Notified by police	1.168	2.93	-	-	-	-
Notified by road ranger	-	-	-	-	0.405	12.23
I-4 road ranger responded	0.584	6.69	-	-	-	-
CFX road ranger responded	0.229	2.62	-	-	-	-
Dependence effects						
Variable		Estimate			t-stat.	
Constant		−1.600			−15.97	
Abandoned activity		−3.309			−3.65	
Debris activity		0.688			5.58	

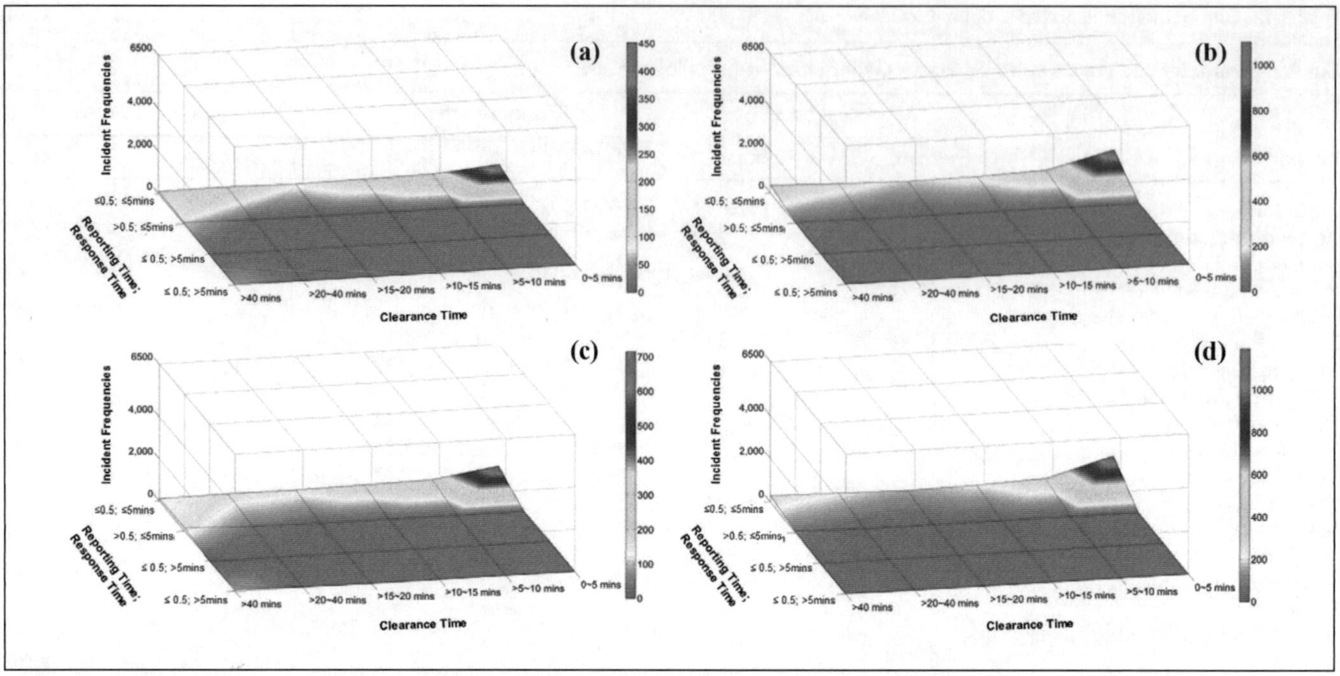

Figure 1. Response category surfaces of predicted events: (*a*) I-4 and morning peak period; (*b*) I-4 and evening peak period; (*c*) roadways other than I-4 and morning peak period; (*d*) roadways other than I-4 and evening peak period.

function of observed exogenous variables in the current study context. The parameter estimates of these components are presented in the third-row panel of Table 4. From Table 4, it can be seen that the variance component of the grouped ordered models is characterized by several exogenous variables. The exogenous variables that contribute to the variance profile of reporting time model include notified by police, I-4 RR responded, and CFX RR responded. The exogenous variables that contribute to the variance profile of response time component of the joint model include at intersection and disabled vehicle. Finally, at rural area, at SR-408, and notified by RR indicator variables are found to have significant effect on the variance profile of clearance time model. Overall, results of these models illustrate the presence and the magnitude of heteroscedasticity in the data.

Dependence Effects. The estimation results of the dependence effects are presented in last row panel of Table 4. Gumbel copula offers an asymmetric dependency structure and the dependency is entirely positive. The coefficient sign and magnitude reflects whether a variable increase or reduces the dependency, and by how much. The dependence results highlight the presence of common unobserved factors affecting different duration components. Further, from the estimated results it can be seen that the copula dependencies are characterized by additional exogenous variables. This provides support to the hypothesis that the dependency structure is not the same across the sample population. The various exogenous variables that contribute to the dependency include crash, abandoned activity, debris activity, tire service, and disabled vehicle.

Model Illustration and Validation Analyses

To demonstrate the implications of the estimated model, the developed model was applied to generate response surface with respect to reporting time, response time, clearance time, and incident frequencies. In generating the values for plotting the response surface, the incident duration categories were identified based on probabilistic assignment by using predicted probabilities computed from the best specified copula model (Gumbel-parameterized). The probabilities are appropriately aggregated across categories to identify the corresponding frequencies. For illustration purposes, the response surfaces are plotted for events on I-4 and other roadways across different time periods (morning peak and evening peak). In Figure 1, predicted incident frequencies (Z-axis) identified based on probabilistic assignments are depicted by three-dimensional charts as a function of the time ranges of reporting/response time and the clearance time categories. In particular, the X-axis includes four outcomes—the combination of two levels of reporting time ($\leqslant$ 0.5 minutes and $>$0.5 minutes) and two levels of response time ($\leqslant$ 5 minutes and $>$ 5 minutes). The Y-axis represents the clearance time. The reader should note that the plots provided are only a sample of the various illustrations that can be generated based on the independent variables in the models. Overall, from the figure it can be observed that incident duration is higher for any time period on I-4 relative to incidents on other roadway locations. Incidents during morning peak result in longer incident durations than incidents in evening peak period, as illustrated in Figure 1. The development of such response surface could be helpful for the incident management agencies to allocate their resources based on the reported incident scenarios.

Conclusion

To understand the overall incident clearance process, this paper formulated and estimated a copula-based trivariate framework accommodating for inherent dependencies across the three components of incident duration: reporting, response, and clearance time. To the best of the authors' knowledge, this is the first attempt to employ a trivariate copula-based methodology in incident duration analysis. Moreover, the study contributes to the incident management literature by examining factors affecting the clearance process, including incident duration components such as incident characteristics, traffic conditions, and roadway and environmental characteristics. The empirical analysis involved estimation of models using four different copula structures: (1) Clayton, (2) Gumbel, (3) Frank, and (4) Joe. The comparison between the copula and the independent models, based on information criterion metrics, confirmed the importance of accommodating dependence across reporting time, response time, and clearance time in incident duration analysis. The most suitable copula model was obtained for Gumbel copula with parameterization for dependence profile. The model estimates were also augmented by conducting policy analysis and 3-dimensional representation of incident frequencies as a function of reporting, response, and clearance time.

The reader should note that the proposed copula approach can be employed to consider hazard duration and/or linear regression-based model structures. However, if the data have a large share of observations concentrated around a single value (nearer to 0 in this study), hazard duration and linear regression models are likely to perform poorly. The proposed approach is not as affected by the large concentrations around a single value. Finally, the proposed model structure does not accommodate for potential endogeneity between the three dependent variables. The consideration for endogeneity is an avenue for future research.

Acknowledgments

The authors would also like to gratefully acknowledge Florida Department of Transportation (FDOT) for providing access to incident data.

Author Contributions

Study conception and design: SY, NE; data collection: HL, SY, NE; analysis and interpretation of results: HL, SY and NE; draft manuscript preparation: HL, SY, NE. All authors reviewed the results and approved the final version of the manuscript.

References

1. Systematics, C. Traffic Congestion and Reliability: Trends and Advanced Strategies for Congestion Mitigation. *Final Report, Texas Transportation Institute*. http://ops. fhwa. dot. gov/congestion_report_04/index. htm, 2005.
2. Roper, D. H. *Freeway Incident Management*. Transportation Research Board, Washington, D.C., 1990.
3. Jodoin, P. *SHRP2 Traffic Incident Management Responder Training*. U.S. Department of Transportation Federal Highway Administration, Washington, D.C., 2016.
4. Ghosh, I., P. Savolainen, and T. Gates. Examination of Factors Affecting Freeway Incident Clearance Times: A Comparison of the Generalized F Model and Several Alternative Nested Models. *Journal of Advanced Transportation*, Vol. 48, 2012, pp. 471–485. https://doi.org/10.1002/atr.
5. *Highway Capacity Manual 2010*. Transportation Research Board of the National Academies, Washington, D.C., 2010.
6. Ma, X., C. Ding, S. Luan, Y. Wang, and Y. Wang. Prioritizing Influential Factors for Freeway Incident Clearance Time Prediction using the Gradient Boosting Decision Trees Method. *IEEE Transactions on Intelligent Transportation Systems*, Vol. 18, No. 9, 2017, pp. 1–8. https://doi.org/10.1109/TITS.2016.2635719.
7. Yin, Y., S. Srinivasan, G. Carrick, K. Jermprapai, X. Sun, N. Shirmohammadi, M. Shahabi, S. Kinney, and A. El-Urfali. *Warrants, Design, and Safety of Road Ranger Service Patrols*. Draft Final Report. The Florida Department of Transportation Research Office, Tallahassee, FL, 2016.
8. Pereira, F. C., F. Rodrigues, and M. Ben-Akiva. Text Analysis in Incident Duration Prediction. *Transportation Research Part C: Emerging Technologies*, Vol. 37, 2013, pp. 177–192. https://doi.org/10.1016/j.trc.2013.10.002.
9. Golob, T. F., W. W. Recker, and J. D. Leonard. An Analysis of the Severity and Incident Duration of Truck-Involved Freeway Accidents. *Accident Analysis and Prevention*, Vol. 19, No. 5, 1987, pp. 375–395. https://doi.org/10.1016/0001-4575(87)90023-6.
10. Giuliano, G.Incident Characteristics, Frequency, and Duration on a High Volume Urban Freeway. *Transportation Research Part A: General*, Vol. 23, No. 5, 1989, pp. 387–396. https://doi.org/10.1016/0191-2607(89)90086-1.
11. Nam, D., and F. Mannering. An Exploratory Hazard-Based Analysis of Highway Incident Duration. *Transportation Research Part A: Policy and Practice*, Vol. 34, No. 2, 2000, pp. 85–102. https://doi.org/10.1016/S0965-8564(98)00065-2.
12. Chimba, D., B. Kutela, G. Ogletree, and F. Horne. Impact of Abandoned and Disabled Vehicles on Freeway Incident Duration. *Journal of Transportation Engineering*, Vol. 140,

No. 3, 2014. https://doi.org/10.1061/(ASCE)TE.1943-5436.0000635.
13. Tavassoli Hojati, A., L. Ferreira, S. Washington, and P. Charles. Hazard Based Models for Freeway Traffic Incident Duration. *Accident Analysis and Prevention*, Vol. 52, 2013, pp. 171–181. https://doi.org/10.1016/j.aap.2012.12.037.
14. Garib, A., A. E. Radwan, and H. Al-Deek. Estimating Magnitude and Duration of Incident Delays. *Journal of Transportation Engineering*, Vol. 123, No. 6, 1997, pp. 459–466. https://doi.org/10.1061/(ASCE)0733-947X(1997)123:6(459).
15. Ding, C., X. Ma, Y. Wang, and Y. Wang. Exploring the Influential Factors in Incident Clearance Time: Disentangling Causation from Self-Selection Bias. *Accident Analysis and Prevention*, Vol. 85, 2015, pp. 58–65. https://doi.org/10.1016/j.aap.2015.08.024.
16. Weng, J., W. Qiao, X. Qu, and X. Yan. Cluster-Based Lognormal Distribution Model for Accident Duration. *Transportmetrica A: Transport Science*, Vol. 11, No. 4, 2015, pp. 345–363. https://doi.org/10.1080/23249935.2014.994687.
17. Khattak, A. J., J. L. Schofer, and M.-H. Wang. A Simple Time Sequential Procedure for Predicting Freeway Incident Duration. *I V H S Journal*, Vol. 2, No. 2, 1995, pp. 113–138. https://doi.org/10.1080/10248079508903820.
18. Valenti, G., M. Lelli, and D. Cucina. A Comparative Study of Models for the Incident Duration Prediction. *European Transport Research Review*, Vol. 2, No. 2, 2010, pp. 103–111. https://doi.org/10.1007/s12544-010-0031-4.
19. Smith, K. W., and B. L. Smith. *Forecasting the Clearance Time of Freeway Accidents*. The Center for Transportation Studies, University of Virginia, Charlottesville, VA, 2001.
20. Demiroluk, S., and K. Ozbay. Structure Learning for the Estimation of Non-Parametric Incident Duration Prediction Models. Presented at 90th Annual Meeting of the Transportation Research Board, Washington, D.C., 2011.
21. Chung, Y. Development of an Accident Duration Prediction Model on the Korean Freeway Systems. *Accident Analysis and Prevention*, Vol. 42, No. 1, 2010, pp. 282–289. https://doi.org/10.1016/j.aap.2009.08.005.
22. Sullivan, E. C. New Model for Predicting Freeway Incidents and Incident Delays. *Journal of Transportation Engineering*, Vol. 123, 1997, pp. 267–275. https://doi.org/10.1061/(ASCE)0733-947X(1997)123:4(267).
23. Hu, J., R. Krishnan, and M. G. H. Bell. Incident Duration Prediction for In-Vehicle Navigation System. Presented at 90th Annual Meeting of the Transportation Research Board, Washington, D.C., 2011.
24. Wang, W., H. Chen, and M. C. Bell. Vehicle Breakdown Duration Modelling. *Journal of Transportation and Statistics*, Vol. 8, No. 1, 2005, pp. 75–84.
25. Ozbay, K., and N. Noyan. Estimation of Incident Clearance Times using Bayesian Networks Approach. *Accident Analysis and Prevention*, Vol. 38, No. 3, 2006, pp. 542–555. https://doi.org/10.1016/j.aap.2005.11.012.
26. Lee, Y., and C. H. Wei. A Computerized Feature Selection Method using Genetic Algorithms to Forecast Freeway Accident Duration Times. *Computer-Aided Civil and Infrastructure Engineering*, Vol. 25, No. 2, 2010, pp. 132–148. https://doi.org/10.1111/j.1467-8667.2009.00626.x.
27. Wu, W., S. Chen, and C. Zheng. Traffic Incident Duration Prediction Based on Support Vector Regression. *Proc., ICCTP 2011: Towards Sustainable Transportation Systems*, Nanjing, China, 2011, pp. 2412–2421.
28. Zhan, C., A. Gan, and M. Hadi. Prediction of Lane Clearance Time of Freeway Incidents using the M5P Tree Algorithm. *IEEE Transactions on Intelligent Transportation Systems*, Vol. 12, No. 4, 2011, pp. 1549–1557. https://doi.org/10.1109/TITS.2011.2161634.
29. Khattak, A. J., X. Wang, and H. Zhang. Incident Management Integration Toll: Dynamically Predicting Incident Durations, Secondary Incident Occurrence, and Incident Delays. *IET Intelligent Transport Systems*, Vol. 6, No. 2, 2012, pp. 204–214. https://doi.org/10.1049/iet-its.2011.0013.
30. Zhang, H., and A. J. Khattak. Analysis of Large-Scale Incidents on Urban Freeways. *Transportation Research Record: Journal of the Transportation Research Board*, 2011. 2178: 30–39.
31. Chung, Y., and B. J. Yoon. Analytical Method to Estimate Accident Duration using Archived Speed Profile and Its Statistical Analysis. *KSCE Journal of Civil Engineering*, Vol. 16, No. 6, 2012, pp. 1064–1070. https://doi.org/10.1007/s12205-012-1632-3.
32. Al Kaabi, A., D. Dissanayake, and R. Bird. Investigating the Response Time of Highway Traffic Accidents in Abu Dhabi using the Hazard-Based Duration Modelling Technique. Presented at 91st Annual Meeting of the Transportation Research Board, Washington, D.C., 2012.
33. Vlahogianni, E. I., and M. G. Karlaftis. Fuzzy-Entropy Neural Network Freeway Incident Duration Modeling with Single and Competing Uncertainties. *Computer-Aided Civil and Infrastructure Engineering*, Vol. 28, No. 6, 2013, pp. 420–433. https://doi.org/10.1111/mice.12010.
34. Tavassoli Hojati, A., L. Ferreira, S. Washington, P. Charles, and A. Shobeirinejad. Modelling Total Duration of Traffic Incidents Including Incident Detection and Recovery Time. *Accident Analysis and Prevention*, Vol. 71, 2014, pp. 296–305. https://doi.org/10.1016/j.aap.2014.06.006.
35. Li, R. Traffic Incident Duration Analysis and Prediction Models Based on the Survival Analysis Approach. *IET Intelligent Transport Systems*, Vol. 9, No. 4, 2015, pp. 351–358. https://doi.org/10.1049/iet-its.2014.0036.
36. Zou, Y., K. Henrickson, D. Lord, Y. Wang, and K. Xu. Application of Finite Mixture Models for Analysing Freeway Incident Clearance Time. *Transportmetrica A: Transport Science*, Vol. 12, No. 2, 2016, pp. 99–115. https://doi.org/10.1080/23249935.2015.1102173.
37. Zhu, W., Z. Zeng, Y. Wang, and Z. Pu. Predicting Incident Duration Based on Spatiotemporal Heterogeneous Pattern Recognition. Presented at 96th Annual Meeting of the Transportation Research Board, Washington, D.C., 2017.
38. Bhat, C. R., and N. Eluru. A Copula-Based Approach to Accommodate Residential Self-Selection Effects in Travel Behavior Modeling. *Transportation Research Part B: Methodological*, Vol. 43, No. 7, 2009, pp. 749–765. https://doi.org/10.1016/j.trb.2009.02.001.
39. Bhat, C. R., and I. N. Sener. A Copula-Based Closed-Form Binary Logit Choice Model for Accommodating Spatial

Correlation across Observational Units. *Journal of Geographical Systems*, Vol. 11, No. 3, 2009, pp. 243–272.
40. Eluru, N., C. R. Bhat, R. M. Pendyala, and K. C. Konduri. A Joint Flexible Econometric Model System of Household Residential Location and Vehicle Fleet Composition/Usage Choices. *Transportation*, Vol. 37, No. 4, 2010, pp. 603–626. https://doi.org/10.1007/s11116-010-9271-3.
41. Zou, Y., and Y. Zhang. A Copula-Based Approach to Accommodate the Dependence among Microscopic Traffic Variables. *Transportation Research Part C: Emerging Technologies*, Vol. 70, 2016, pp. 53–68.
42. Trivedi, P. K., and D. M. Zimmer. Copula Modeling: An Introduction for Practitioners. *Foundations and Trends® in Econometrics*, Vol. 1, No. 1, 2007, pp. 1–111.
43. Sklar, A. Random Variables, Joint Distribution Functions, and Copulas. *Kybernetika*, Vol. 9, No. 6, 1973, pp. 449–460.
44. Yasmin, S., N. Eluru, A. R. Pinjari, and R. Tay. Examining Driver Injury Severity in Two Vehicle Crashes–A Copula Based Approach. *Accident Analysis & Prevention*, Vol. 66, 2014, pp. 120–135.
45. Wang, K., S. Yasmin, K. C. Konduri, N. Eluru, and J. N. Ivan. Copula-Based Joint Model of Injury Severity and Vehicle Damage in Two-Vehicle Crashes. *Transportation Research Record: Journal of the Transportation Research Board*, 2015. 2514: 158–166.
46. Eluru, N., R. Paleti, R. Pendyala, and C. Bhat. Modeling Injury Severity of Multiple Occupants of Vehicles: Copula-Based Multivariate Approach. *Transportation Research Record: Journal of the Transportation Research Board*, 2010. 2165: 1–11.
47. Eluru, N., I. Sener, C. Bhat, R. Pendyala, and K. Axhausen. Understanding Residential Mobility: Joint Model of the Reason for Residential Relocation and Stay Duration. *Transportation Research Record: Journal of the Transportation Research Board*, 2009. 2133: 64–74.
48. Aptech. GAUSS, Aptech Systems, Inc. http://www.aptech.com/. Accessed September 19, 2015.
49. Hou, L., Y. Lao, Y. Wang, Z. Zhang, Y. Zhang, and Z. Li. Modeling Freeway Incident Response Time: A Mechanism-Based Approach. *Transportation Research Part C: Emerging Technologies*, Vol. 28, 2013, pp. 87–100. https://doi.org/10.1016/j.trc.2012.12.005.
50. Hou, L., Y. Lao, Y. Wang, Z. Zhang, Y. Zhang, and Z. Li. Time-Varying Effects of Influential Factors on Incident Clearance Time using a Non-Proportional Hazard-Based Model. *Transportation Research Part A: Policy and Practice*, Vol. 63, 2014, pp. 12–24. https://doi.org/10.1016/j.tra.2014.02.014.
51. Florida Department of Highway Safety and Motor Vehicles. *Traffic Crash Facts: Annual Report 2015*, 2016.

The Standing Committee on Statistical Methods (ABJ80) peer-reviewed this paper (18-03310).

Article

TRR
JOURNAL OF THE TRANSPORTATION RESEARCH BOARD

Transportation Research Record
2018, Vol. 2672(30) 90–98
© National Academy of Sciences:
Transportation Research Board 2018
Article reuse guidelines:
sagepub.com/journals-permissions
DOI: 10.1177/0361198118782249
journals.sagepub.com/home/trr
SAGE

Jointly Specified Spatial Priors for Bayesian Models of Crash Frequency

Jonathan Aguero-Valverde[1]

Abstract
The presence of spatial correlation in crash frequency data has being well documented. Most of the studies in highway safety used conditionally specified spatial errors, such as Conditional Autoregressive (CAR) random effects. An alternative to conditionally specified models is to choose directly a parametric functional form for the covariance matrix of the error term, also known as jointly specified models of spatial correlation. The main advantage of these models is that they do not depend on an arbitrary defined proximity matrix. The purpose of the study is to demonstrate the use of jointly specified spatial priors for Bayesian models of crash frequency using four types of correlation functions: exponential, rational quadratic, Gaussian, and Matérn. The frequency of equivalent-property-damage-only crashes in 162 intersections on Route 34 in Costa Rica was modeled. The results show that the four jointly specified models are similar in goodness-of-fit and significantly better than the CAR model. The estimated mean range varies from about 20 m to more than 60 m and the mean effective range varies between 100 m and 200 m. Most of the information on the posterior of the range comes from the data, since the posteriors diverge significantly from the uniform priors. There is not enough evidence to select a correlation function over the other three; further studies are, therefore, recommended to shed light on this issue.

Several studies have demonstrated the presence of spatial correlation in crash frequency data (*1–16*). These analyses range from intersections (*4, 8, 12, 16*) and segments (*5, 7, 12–15*) to areal units such as counties (*1, 3, 9, 11*), census wards (*6*), Traffic Analysis Zones (TAZ) (*10*), and local health areas (*2*). Ignoring spatial correlation can lead to biased or inefficient estimates (*17–19*). Spatial models have, therefore, been introduced in highway safety to improve model estimation, particularly in the presence of small sample sizes and low sample means.

Classical regression methods are only valid when the residuals have no spatial structure since they are assumed to be independent. When spatial correlation is present, estimation should proceed with a spatially structured covariance error matrix (*20*). Most of the studies in highway safety used conditionally specified (Markov random field) models to specify the covariance error matrix. Spatial models such as conditional autoregressive (CAR) (*21, 22*) or spatial autoregressive (SAR) (*23*) are examples of conditional specified structures. These models express spatial dependence in terms of a previously defined proximity or adjacency matrix. CAR models are the most commonly used spatial correlation models in highway safety (*1, 3, 5–7, 9–12, 14–16*).

An alternative to conditionally specified models is to choose directly a parametric functional form for the covariance matrix of the error term (*24*). These jointly specified models of spatial correlation have been used in disease mapping (*20, 24–29*) but not in highway safety with Aguero-Valverde being probably the only exception (*13*). The main advantage of these models is that they do not depend on an arbitrary defined proximity matrix and the strength of the spatial correlation is obtained from the parameters of the correlation function. One drawback of jointly specified models in disease mapping is the fact that each area has to be represented as a point to measure the distances between spatial units; frequently, the centroid is used for this purpose. In road segments, the issue is similar and the center of the segment can be used (*13*). For intersections, however, this is not an issue since they are basically nodes (i.e., points) in the road network and this fits easily onto distance-based modeling.

There are few comparisons of jointly specified and CAR models in the literature. The only previous example in crash frequency modeling showed a significant

[1]Programa de Investigación en Desarrollo Urbano Sostenible, Universidad de Costa Rica, Barrio Los Profesores, San José, Costa Rica

Corresponding Author:
Address correspondence to Jonathan Aguero-Valverde: jonathan.aguero@ucr.ac.cr

improvement for the exponential correlation model compared with the CAR model (*13*). It is important to note that this was a segment model, not an intersection model and the distance used was Cartesian distance (i.e., the shortest distance between two points) rather than network distance (i.e., distance driven). On the other hand, Best et al. (*24*) found that the CAR model performed better than the exponential correlation model in a simulation study of disease risk at the census enumeration district level in England. Wall (*29*) also found the CAR model to fit the data on statewide verbal scores on the SAT better than the exponential covariogram model.

The purpose of this paper is to demonstrate the use of jointly specified spatial priors for Bayesian models of crash frequency using different types of correlation. Four jointly specified spatial models are estimated using different correlation functions: exponential, rational quadratic, Gaussian, and Matérn. A model with CAR error terms is also estimated for comparison purposes. This paper is organized as follows: first the different spatial structures used in the paper are presented in the methods section. The description of the data used in the analysis is presented in the next section. Then, the results section presents a comparative analysis between representative spatial models. The article concludes with a discussion and suggestions for future work.

Methodology

At the first level of the hierarchy a Poisson distribution is proposed to model the number of crashes at each intersection:

$$Y_i \sim Poisson(\theta_i) \tag{1}$$

where Y_i is the observed number of crashes at intersection *i*, and θ_i is the expected Poisson rate for intersection *i*. The Poisson rate is modeled as a function of the covariates with a log-link function as shown is Equation 2:

$$\log(\theta_i) = \beta_0 + \sum_{k=1}^{K} \beta_k x_{ik} + s_i \tag{2}$$

where β_0 is the intercept, β_k is the coefficient for *k* covariate, x_{ik} is the value of the k covariate for intersection *i*, and s_i is the spatially correlated random error. This random error captures the extra-Poisson heterogeneity among intersections.

At the next stage, the coefficients (β_k), including the intercepts, are modeled using very non-informative Normal priors:

$$\beta_k \sim N(0, 1000) \tag{3}$$

The multivariate normal distribution is a flexible distribution often used for representing correlated random errors. If $\boldsymbol{S} = \{s_1, \ldots, s_n\}$ denotes the vector of intersection specific random errors in Equation 2, the dependence structure can be specified in terms of an $n \times n$ covariance matrix $\boldsymbol{\Sigma}$, leading to a second stage prior:

$$S \sim \mathbf{MVN}(\mu, \boldsymbol{\Sigma}) \tag{4}$$

where $\boldsymbol{\mu}$ is a vector of zeroes, $\boldsymbol{\Sigma} = \sigma^2\boldsymbol{\Omega}$ and Ω_{ij} is the correlation between s_i and s_j. In the case of this study, the elements of the correlation matrix are specified as a function of the road network distance d_{ij} between each pair of intersections:

$$\Omega_{ij} = f(d_{ij}; \phi) \tag{5}$$

where $\boldsymbol{\phi}$ is the vector of parameters of the correlation function. The correlation function needs to be carefully selected to ensure that the function results in a positive definite covariance matrix $\boldsymbol{\Sigma}$. One of the most widely used correlation function is the exponential decay function (Best et al. 2005):

$$\Omega_{ij} = e^{\frac{-d_{ij}}{h}} \tag{6}$$

where the parameter $h > 0$ is known as the range. The range controls the rate of decrease of the correlation with distance.

Other correlation functions can be implemented. The rational quadratic function is another one-parameter function:

$$\Omega_{ij} = 1 - \frac{d_{ij}^2}{1 + \frac{d_{ij}^2}{h^2}} \tag{7}$$

Different shapes of the decrease with the distance can be modeled using two parameter functions such as the power exponential family or the Matérn class (*24*). A special case of the power exponential family is the Gaussian model where the exponent is equal to 2:

$$\Omega_{ij} = e^{\frac{-d_{ij}^2}{h^2}} \tag{8}$$

Note that the exponential function also belongs to the power exponential family.

The Matérn class of correlation functions was also explored. For this study, the closed form of the Matérn function at $\nu = 3/2$ was implemented:

$$\Omega_{ij} = \left(1 + \frac{d_{ij}}{h}\right) e^{\frac{-d_{ij}}{h}} \tag{9}$$

Table 1. Definition of the Effective Range for Each Correlation Function

Correlation model	Effective range (d_0)
Exponential	2.99573227*h
Rational quadratic	4.3588*h
Gaussian	1.73081838*h
Matérn (ν = 3/2)	4.7439*h

Interested readers can find more details about the Matérn function and the other correlation functions used in this paper in Banerjee et al. (*30*).

The effective range, that is, the distance at which there is essentially no lingering spatial correlation, is also of interest. A common definition of this measure is the distance at which the correlation has dropped to only 5% (*30*). Consequently, the effective range d_0 was estimated for each correlation function used in this paper as shown in Table 1.

For comparison purposes, a CAR model was also estimated. A CAR distribution can be thought of as normally conditional on its neighbors:

$$s_i|s_{-i} \sim \mathrm{N}\left(\frac{\sum_j w_{ij}s_{j\sim i}}{w_{i+}}, \frac{1}{\tau w_{i+}}\right) \quad (10)$$

where s_{-i} are all the neighbors of site i, $j \sim i$ denotes that site j is a neighbor of site i, w_{ij} is the weight of the jth neighbor of the ith site, w_{i+} is the sum of the weights of the neighbors of site i and τ controls the Poisson extra-variation due to spatial correlation.

The variance at zero distance (σ^2), also known as sill in geostatistics (*18*), was modeled with a semi-informative Gamma (0.1, 0.1) distribution as prior. For the range, uniform priors were used. In the case of the exponential and Matérn functions, the prior for the range was h˜uniform (0, 2000). Since the distance is measured in meters, the upper limit of the prior distribution is 2 km, which was considered a reasonable prior expectation. For the rational quadratic and Gaussian models the prior selected was $h^2 \sim$ uniform (0, 4000000). The upper limit here corresponds to the square of 2 km. These priors are selected with a relatively small upper limit to avoid nonidentifiability of the correlation parameters.

Two different goodness-of-fit (GOF) measures frequently used in Bayesian statistics are used in this study: posterior mean deviance ($\bar{D}$) and Deviance Information Criterion (DIC). The posterior mean deviance can be taught as a Bayesian measure of fit or 'adequacy.' To account for model complexity the Deviance Information Criterion was proposed by Spiegelhalter et al. (*31*). The DIC is considered the Bayesian equivalent of the Akaike Information Criterion (AIC). DIC is defined as an estimate of fit plus twice the effective number of parameters:

$$\mathrm{DIC} = \mathrm{D}(\bar{\theta}) + 2p_D = \bar{D} + p_D \quad (11)$$

Where $D(\bar{\theta})$ is the deviance evaluated at $\bar{\theta}$ the posterior means of the parameters of interest, p_D is the effective number of parameters in the model, and $\bar{D}$ is the posterior mean of the deviance statistic $D(\theta)$. As with AIC, models with lower DIC values are preferred.

Models are estimated using OpenBUGS 3.2.3 (*31*). The first five thousand iterations are discarded as burn-in for each model. After that, 100,000 iterations are saved to produce summary statistics for each parameter in the model. Monte Carlo (MC) errors in the parameter estimates are lower than 5% of the standard deviation of the parameter for all the estimates; this ensures that the variability in the parameter is twenty times higher than the measurement error (MC error).

Data Description

The crash dataset was obtained from COSEVI (Costa Rican Highway Safety Council). Crash frequencies in each intersection are aggregated from 2011 to 2013. The data correspond to all the intersections in Route 34 which travels from the Central region of Costa Rica to the South Pacific region. Route 34 is more than 200 km in length and has 162 intersections. To incorporate the severity of each crash into the analysis, crashes were weighted and transformed to equivalent-property-damage-only crashes using the weights of 40, 12, and 3 for fatal, major injury, and minor injury respectively as suggested by Castro and Aguero-Valverde (*32*).

Road network characteristics were obtained from the National Highway Inventory maintained by the National Department of Transportation. Unfortunately, most of the minor roads are local roads for which inventory data are not available; for this reason no annual average daily traffic (AADT) data for minor roads exist. To mitigate this issue, the type of minor road was introduced into the model to control, at least in part, for the traffic in the secondary roads at the intersections. Table 2 presents the descriptive statistics of the data used in the models.

AADT on the primary roads ranges from 6,408 to more than 50,000 and only 5.6% of the minor roads are national roads. In terms of land use, 66% of the intersections are located in an urban environment while 85% of them are 3-legged intersections. Also, most crashes in the study region only caused property damage. During the study period only two fatal crashes, four major injury crashes, and 29 minor injury crashes, out of 266 total crashes, were reported.

Table 2. Descriptive Statistics of the Intersections on Route 34, Costa Rica

Variable	Mean	Standard deviation	Minimum	Maximum
Fatal Crashes	0.012	0.111	0	1
Major Injury Crashes	0.025	0.156	0	1
Minor Injury Crashes	0.179	0.485	0	3
Property Damage Only Crashes	1.642	3.555	0	33
Total Crashes	1.858	3.812	0	35
Equivalent-property-damage-only crashes	2.969	6.974	0	50
AADT Primary Road	18799.7	8302.9	6408	50518
Categorical variables	Percentage			
Type of minor road: national	5.6			
Type of minor road: local	94.4			
Urban zone	66.0			
Intersection Type: 4-legged	11.7			
Intersection Type: 3-legged	85.8			
Intersection Type: Other	2.5			

Results

Table 3 shows the coefficient estimates and goodness-of-fit measures for the jointly specified spatial models as well as the CAR model. Spiegelhalter et al. (*33*) suggest that models with DIC values within 1 or 2 points of the "best" model are strongly supported, those with values between 3 and 7 points of the "best" model are only weakly supported and models with a DIC more than 7 points higher than the best model are substantially inferior. As shown in Table 3, the model with the lowest DIC and posterior deviance is the exponential model; however, the other 3 models with covariance functions are within a point of the "best" model. On the other hand, the CAR model presents a DIC of 512.1 or more than 7 points higher than the jointly specified models. These results suggest that there is almost no difference among the models with jointly specified spatial priors but they are substantially better than the CAR model for the data used in this study. These results are consistent with a previous study that showed a significant improvement of 72 points in DIC for the exponential correlation model compared with the CAR model (*13*). Interestingly, models based on areas rather than road networks seem to favor CAR over jointly specified models (*24, 29*).

The effective number of parameters is p_D small relative to the number of data points which indicates important borrowing of strength across intersections. This number is about 95 for the jointly specified models while the number of intersections is 162. On the other hand, p_D. for the CAR model is 85 which implies that the CAR prior contains more structural information than the jointly specified spatial priors.

The mean range varies in the correlation models from about 20 m in the Matérn model to more than 60 m for the Gaussian model. More importantly, the mean effective range (d_0) varies from 99.5 m for the Matérn model to more than 207 m for the rational quadratic model. Interestingly, these values are very similar to the one obtained by Aguero-Valverde (*13*) that found a value for the effective range of 168.7 m for an exponential model of crash frequency at segment level. To further explore these variables their posterior distributions are shown in Figure 1. It is clear from the distributions that the variables are right-skewed as you might expect for a variable that is strictly positive. Skewed distributions are better represented by the median instead of the mean; the mean and the median for these variables are reported in Table 4. The exponential and Matérn models show more skewness than the rational quadratic and Gaussian covariograms. In terms of centrality measures, the exponential model shows the biggest differences between the mean and the median with differences of about 5 m and 20 m for the range and effective range respectively.

The effect of the prior distributions on the posterior of the covariogram parameters is of interest, given the small sample size. The priors for h and h^2 are uniform but with a relatively small upper limit of 2 km to avoid nonidentifiability of the correlation parameters. However, most of the information on the posterior of the parameters comes from the data, as the posteriors diverge significantly from the uniform priors.

The variances are very similar in all the correlation models, varying from 2.35 for the Matérn model to 2.41 for the Gaussian function. The CAR model explains less variance with a σ^2 (inverse of τ in Equation 10) of 2.24.

Figure 2 presents the estimated covariograms. A covariogram can be seen as the product of the variance at zero distance or sill (σ^2) times the correlation function; the covariograms, therefore, present the complete effect of the spatial correlation over distance. The similarities

Table 3. Jointly Specified Spatial and CAR Models

Model	Exponential				Rational quadratic				Gaussian			
	Mean	SD	$P_{2.5}$	$P_{97.5}$	Mean	SD	$P_{2.5}$	$P_{97.5}$	Mean	SD	$P_{2.5}$	$P_{97.5}$
Constant	−5.637	3.662	−12.570	1.731	−5.723	3.439	−12.470	1.020	−5.526	3.666	−12.590	1.643
Ln (AADT major road)	0.816	0.371	0.079	1.519	0.825	0.350	0.136	1.515	0.802	0.376	0.064	1.528
Minor road type: Local	−1.429	0.648	−2.727	−0.172	−1.355	0.669	−2.681	−0.016	−1.406	0.671	−2.757	−0.160
3-legged intersection	−1.407	0.427	−2.242	−0.565	−1.481	0.457	−2.410	−0.627	−1.413	0.434	−2.269	−0.566
other intersection	−0.073	1.004	−2.006	1.966	−0.037	0.992	−1.938	1.969	0.027	1.010	−2.021	1.966
σ^2	2.375	0.528	1.530	3.583	2.380	0.525	1.524	3.570	2.409	0.539	1.550	3.647
h	51.5	35.6	2.6	136.5	47.7	22.4	10.0	96.5	60.6	25.5	12.8	111.2
h^2					2777.0	2427.0	100.5	9319.0	4316.0	3268.0	163.7	12360.0
d_0	154.4	106.7	7.9	409.0	207.8	97.8	43.7	420.8	104.8	44.1	22.2	192.4
$\bar{D}$	409.7				410.5				410.2			
$\hat{D}$	315.0				316.1				315.7			
p_D	94.6				94.5				94.6			
DIC	504.3				505.0				504.8			

Model	Matérn				CAR			
	Mean	SD	$P_{2.5}$	$P_{97.5}$	Mean	SD	$P_{2.5}$	$P_{97.5}$
Constant	−5.585	3.448	−12.420	1.135	−5.357	10.920	−27.510	14.930
Ln (AADT major road)	0.818	0.353	0.128	1.517	0.806	1.123	−1.299	3.082
Minor road type: Local	−1.454	0.637	−2.742	−0.249	−0.951	0.544	−2.044	0.101
3-legged intersection	−1.456	0.437	−2.328	−0.595	−2.302	0.456	−3.222	−1.424
other intersection	−0.113	0.919	−1.981	1.647	−0.266	0.939	−2.102	1.584
σ^2	2.349	0.514	1.514	3.513	2.239	0.560	1.352	3.531
h	21.0	13.4	1.0	50.0				
h^2								
d_0	99.5	63.5	4.8	237.1				
$\bar{D}$	410.2				427.1			
$\hat{D}$	315.6				342.1			
p_D	94.6				85.0			
DIC	504.8				512.1			

Table 4. Mean and Median for Estimated Range of Effective Range

Model	Range (h)		Effective range (d_0)	
	Mean	Median	Mean	Median
Exponential	51.53	45.88	154.4	137.4
Rational quadratic	47.68	45.77	207.8	199.5
Gaussian	60.55	60.86	104.8	105.3
Matérn	20.96	19.8	99.45	93.95

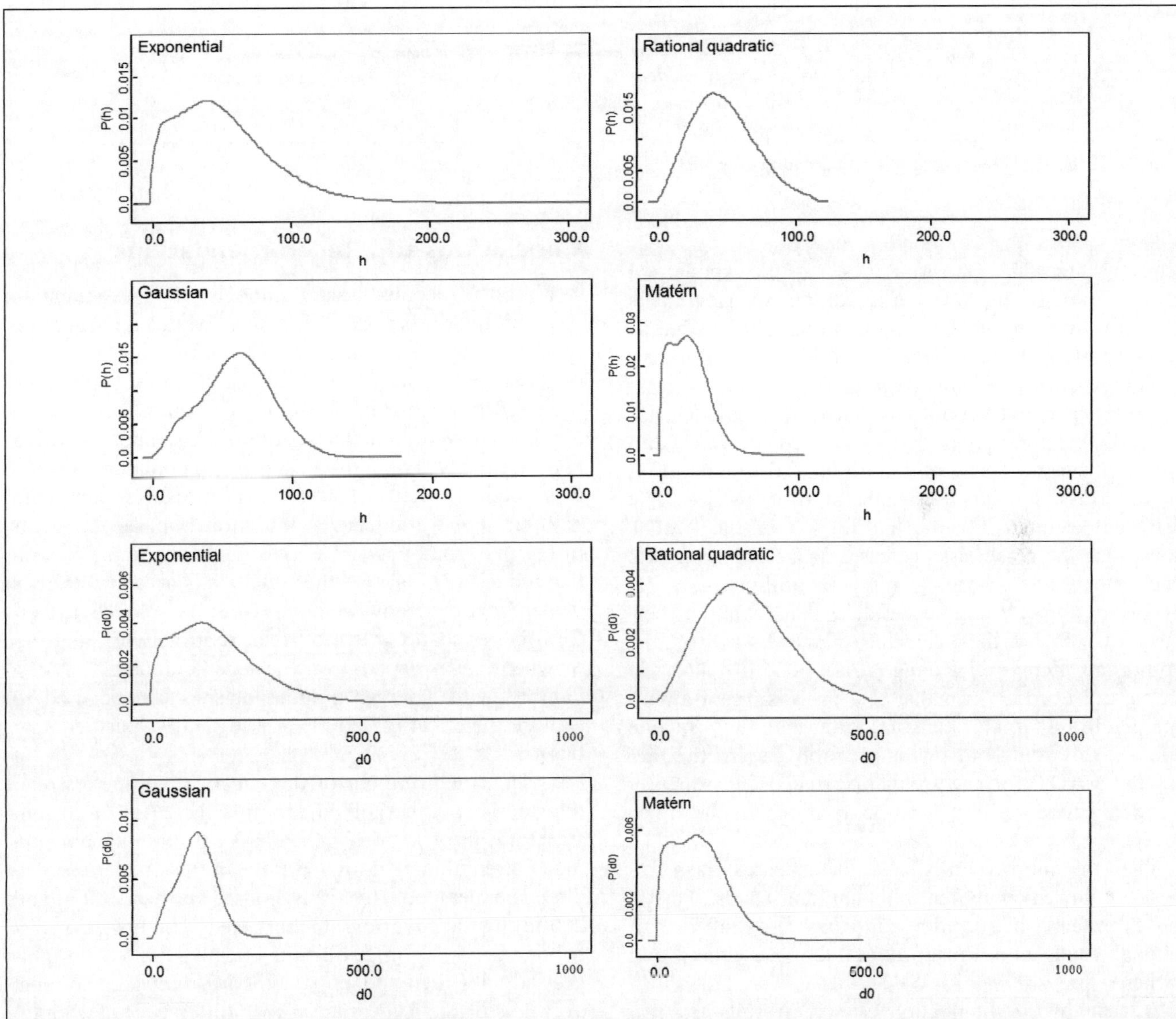

Figure 1. Posterior distribution of the range (h) and the "effective range" (d_0).

between the four covariograms support the previous finding that there are no detectable differences between the correlation functions. However, there are some differences to highlight in the models. The effective range, or 5% correlation threshold, is reached faster by the Gaussian and Matérn covariograms at about 100 m. Also, the exponential covariogram decreases faster than all the other functions until about half of the correlation,

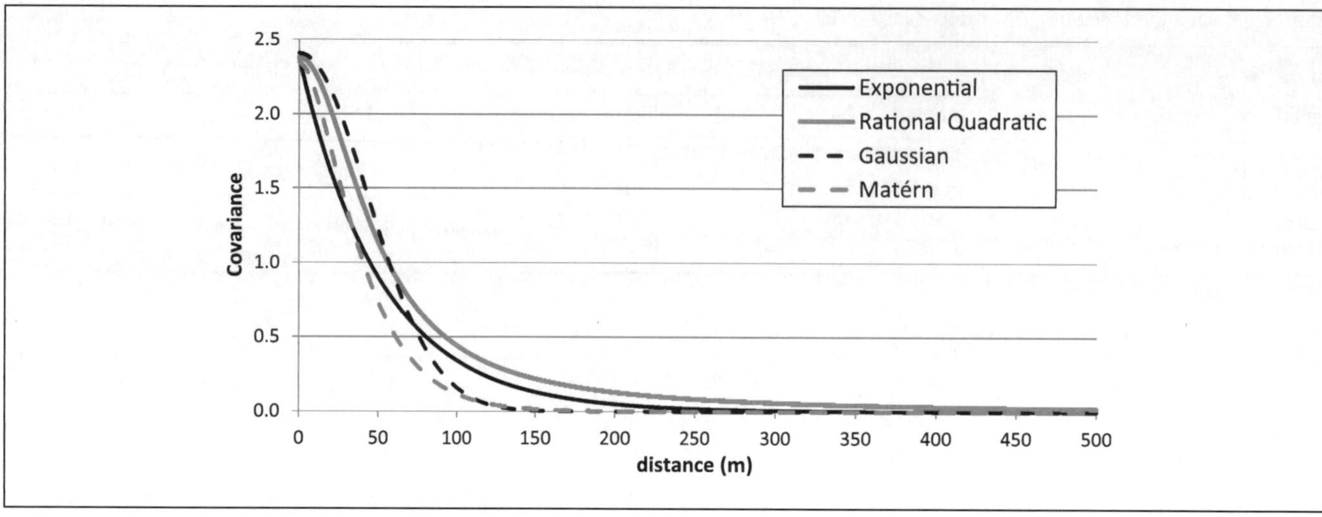

Figure 2. Estimated covariograms for different models.

then the Matérn and Gaussian lines cross the exponential function. Finally, the rational quadratic covariogram is in general over the other functions which implies that at most distances the correlation in this model is higher than in the other three covariograms.

The independent variables included in the models are the AADT of the major road, type of minor road, and type of intersection. The land use around the intersection was found not statistically significant in the models. The coefficient estimate for the natural log of the AADT shows that the crash risk increases at a decreasing rate which is expected according to several authors (*4, 8, 12, 34*). For example, Wang and Abdel-Aty (*4*) found a coefficient of 0.56 for the logarithm of the AADT on the major road, Zeng and Huang (*12*) a value of 0.46, and Miaou and Lord (*34*) a value of 0.54. It is important to note that those sample values are lower than the 0.8 value found in this study but this is probably due to the fact that the AADT for the secondary road is not directly considered here while it was incorporated in the other models.

When the minor road is local, the expected crash frequency is lower compared with national roads. This is also expected, since local roads carry less traffic than national roads and thus produce fewer conflicts and crashes.

In terms of the intersection type, the results are also consistent with previous research that found that 3-legged intersections are safer than 4-legged intersections (*35–37*). In the present study the coefficient is around –1.4 while comparable studies reported values varying from –0.66 to –1.048 (*36*), or –0.18 to –0.29 (*37*). This is clearly expected since 3-legged intersections have fewer conflict points and, as a consequence, less potential for crashes.

Conclusions and Recommendations

The jointly specified spatial models using exponential, rational quadratic, Gaussian and Matérn correlation functions show little differences in terms of goodness-of-fit measures but all of them are significantly better than the CAR model for crash frequency at intersection level for the analyzed data. These results are consistent with a previous study that showed that the exponential correlation model presents better GOF measures compared with the CAR model (*13*) but models based on areas rather than road networks favor CAR over jointly specified models (*24, 29*). Further research is needed to clarify if there is a tendency for jointly specified models to perform better than CAR models in road network analysis or whether this depends on the specific dataset. Besides, datasets with a larger number of sites can be used to explore the effect of sample size in jointly specified spatial models.

The small number of the effective parameters relative to the number of data points indicates important structural information in the spatial priors and borrowing of strength across intersections in the study area.

The mean effective range varies between 100 m and 200 m for the covariograms modeled. The posterior distributions of the range and the effective range are right-skewed with the exponential function showing the most skewness of the four models, with differences of about 5 m and 20 m between the mean and the median for the range and the effective range respectively.

The priors have little effect on the posterior distribution of the covariance parameters. Most of the information on the posterior of the parameters comes from the data, since the posteriors diverge significantly from the uniform priors.

The value of the variance is very close in all the jointly specified models which also shows that these models are very similar. This fact is also reinforced by the covariograms that present many similarities. In summary, there is not enough evidence to select a correlation function over the other three. Further study is, therefore, recommended to shed light on this issue.

References

1. Miaou, S.-P., J. J. Song, and B. K. Mallick. Roadway Traffic Crash Mapping: A Space-time Modeling Approach. *Journal of Transportation and Statistics*, Vol. 6, No 1, 2003, pp. 33–57.
2. MacNab, Y. C. Bayesian Spatial and Ecological Models for Small-area Accident and Injury Analysis. *Accident Analysis and Prevention*, Vol. 36, No. 6, 2004, pp. 1091–1028.
3. Aguero-Valverde, J., and P. P. Jovanis. Spatial Analysis of Fatal and Injury Crashes in Pennsylvania. *Accident Analysis and Prevention*, Vol. 38, No. 3, 2006, pp. 618–625.
4. Wang, X., and M. Abdel-Aty. Temporal and Spatial Analysis of Rear-end Crashes at Signalized Intersections. *Accident Analysis and Prevention*, Vol. 38, No. 6, 2006, pp. 1137–1150.
5. Aguero-Valverde, J., and P. P. Jovanis. Analysis of Road Crash Frequency with Spatial Models. *Transportation Research Record: Journal of the Transportation Research Board*, 2008. 2061: 55–63.
6. Quddus, M. A. Modeling Area-wide Count Outcomes with Spatial Correlation and Heterogeneity: An Analysis of London Crash Data. *Accident Analysis and Prevention*, Vol. 40, 2008, pp. 1486–1497.
7. Aguero-Valverde, J., and P. P. Jovanis. Spatial Correlation in Multilevel Crash Frequency Models: Effects of Different Neighboring Structures. *Transportation Research Record: Journal of the Transportation Research Board*, 2010. 2165: 21–33.
8. Guoa, F., X. Wang, and M. A. Abdel-Aty. Modeling Signalized Intersection Safety with Corridor-level Spatial Correlations. *Accident Analysis & Prevention*, Vol. 42, No. 1, 2010, pp. 84–92.
9. Huang, H., A. L., Darwiche, and M. A. Abdel-Aty. County-Level Crash Risk Analysis in Florida: Bayesian Spatial Modeling. *Transportation Research Record: Journal of the Transportation Research Board*, 2010. 2148: 27–37.
10. Siddiqui, C., M. A Abdel-Aty, and K. Choi. Macroscopic Spatial Analysis of Pedestrian and Bicycle Crashes. *Accident Analysis & Prevention*, Vol. 45, 2012, pp. 382–391.
11. Aguero-Valverde, J. Multivariate Spatial Models of Excess Crash Frequency at Area Level: Case of Costa Rica. *Accident Analysis &* Prevention, Vol. 59, 2013, pp. 365–373.
12. Zeng, Q., and H. Huang. Bayesian Spatial Joint Modeling of Traffic Crashes on an Urban Road Network. *Accident Analysis & Prevention*, Vol. 67, 2014, pp. 105–112.
13. Aguero-Valverde, J. Direct Spatial Correlation in Crash Frequency Models: Estimation of the Effective Range. *Journal of Transportation Safety & Security*, Vol. 6, No. 1, 2014, pp. 21–33.
14. Barua, S., K. El-Basyouny, and M. T. Islam. A Full Bayesian Multivariate Count Data Model of Collision Severity with Spatial Correlation. *Analytic Methods in Accident Research*, Vol. 3, 2014, pp. 28–43.
15. Aguero-Valverde, J., K.-F. Wu, and E. T. Donnell. A Multivariate Spatial Crash Frequency Model for Identifying Sites with Promise Based on Crash Types. *Accident Analysis & Prevention*, Vol. 87, 2016, pp. 8–16.
16. Guadamuz-Flores, R., and J. Aguero-Valverde. Bayesian Spatial Models of Crash Frequency at Highway-Railway Crossings. *Transportation Research Record: Journal of the Transportation Research Board*, 2017. 2608: 27–35.
17. Dubin, R. A. Estimation of Regression Coefficients in the Presence of Spatially Autocorrelated Error Terms. *Review of Economics and Statistics*, Vol. 70, No. 3, 1988, pp. 466–474.
18. Cressie, N. A. C. *Statistics for Spatial Data*. John Wiley & Sons, New York, 1993.
19. Congdon, P. *Bayesian Statistical Modeling*. John Wiley & Sons, England, 2001.
20. Richardson, S. Statistical Methods for Geographical Correlation Studies. In *Geographical and Environmental Epidemiology* (P. Elliott, J. Cuzick, D. English, and R. Stern eds.), Oxford University Press, 1992, pp. 181–204.
21. Besag, J. Spatial Interaction and the Statistical Analysis of Lattice Systems. *Journal of the Royal Statistical Society, Series B*, Vol. 36, No. 2, 1974, pp. 192–236.
22. Besag, J., J. C. York, and A. Mollié. Bayesian Image Restoration, with Two Applications in Spatial Statistics. *Annals of the Institute of Statistical Mathematics*, Vol. 43, No. 1, 1991, pp. 1–59.
23. Whittle, P. On Stationary Processes in the Plane. *Biometrika*, Vol. 41, No. 4-5, 1954, pp. 434–449.
24. Best, N., S. Richardson, and A. Thomson. A Comparison of Bayesian Spatial Models for Disease Mapping. *Statistical Methods in Medical Research*, Vol. 14, pp. 35–59.
25. Cook, D. G., and S. J. Pocock. Multiple Regression in Geographic Mortality Studies with Allowance for Spatially Correlated Errors. *Biometrics*, Vol. 39, No. 2, 1983, pp. 361–371.
26. Richardson, S., C. Guihenneuc, and V. Lasserre. Spatial Linear Models with Autocorrelated Error Structure. *The Statistician*, Vol. 41, No. 5, 1992, pp. 539–557.
27. Diggle, P. J., J. A. Tawn, and R. A. Moyeed. Model Based Geostatistics. *Journal of the Royal Statistical Society Series C (Applied Statistics)*, Vol. 47, No. 3, 1998, pp. 299–350.
28. Wakefield, J. C., N. G. Best, and L. A. Waller. Bayesian Approaches to Disease Mapping. In *Spatial Epidemiology: Methods and Applications* (P. Elliott, J. C. Wakefield, N. G. Best, and D. J. Briggs eds.), Oxford University Press, Oxford, 2000, pp. 104–127.
29. Wall, M. M. A Close Look at the Spatial Structure Implied by the CAR and SAR Models. *Journal of Statistical Planning and Inference*. Vol. 121, No. 2, 2004, pp. 311–324.
30. Banerjee, S., B. P. Carlin, and A. E. Gelfand. *Hierarchical Modeling and Analysis for Spatial Data*. Chapman & Hall, 2004.
31. Lunn, D., D. Spiegelhalter, A. Thomas, and N. Best. The BUGS Project: Evolution, Critique and Future Directions.

Statistics in Medicine, Vol. 28, No. 25, 2009, pp. 3049–3067.

32. Castro, E. A., and J. Aguero-Valverde. Aplicación del Método Bayes Empírico en Análisis de Seguridad Vial: El Caso de la Ruta 32 en Costa Rica. *Journal of Tropical Engineering*, Vol. 25, No. 1, 2015, pp. 13–33.
33. Spiegelhalter, D., N. Best, B. P. Carlin, and A. Linde. Bayesian Measures of Model Complexity and Fit. *Journal of the Royal Statistical Society B*, Vol. 64, No. 4, 2002, pp. 583–639.
34. Miaou, S. P., and D. Lord. Modeling Traffic Crash-flow Relationships for Intersections: Dispersion Parameter, Functional Form, and Bayes Versus Empirical Bayes Methods. *Transportation Research Record: Journal of the Transportation Research Board*, 2003. 1840: 31–40.
35. Haque, M. M, H. C. Chin, and H. Huang. Applying Bayesian Hierarchical Models to Examine Motorcycle Crashes at Signalized Intersections. *Accident Analysis & Prevention*, Vol. 42, No. 1, 2010, pp. 203–212.
36. Castro, M., R. Paleti, and C. R. Bhat. A Latent Variable Representation of Count Data Models to Accommodate Spatial and Temporal Dependence: Application to Predicting Crash Frequency at Intersections. *Transportation Research Part B: Methodological*, Vol. 46, No. 1, 2012, pp. 253–272.
37. Xie, K., X. Wang, H. Huang, and X. Chen. Corridor-level Signalized Intersection Safety Analysis in Shanghai, China Using Bayesian Hierarchical Models. *Accident Analysis & Prevention*, Vol. 50, 2013, pp. 25–33.

The Standing Committee on Statistical Methods (ABJ80) peer-reviewed this paper (18-05317).

Article

TRR
JOURNAL OF THE TRANSPORTATION RESEARCH BOARD

Transportation Research Record
2018, Vol. 2672(30) 99–108
© National Academy of Sciences:
Transportation Research Board 2018
Article reuse guidelines:
sagepub.com/journals-permissions
DOI: 10.1177/0361198118788205
journals.sagepub.com/home/trr
SAGE

Highway Safety Manual Calibration: Assessing Alternate Definitions of the Calibration Factor

Mahdi Rajabi[1], Jennifer Harper Ogle[1], and Patrick Gerard[2]

Abstract

The publication of the *Highway Safety Manual* (HSM) in 2010 established crash frequency prediction as the essential safety measure for safety studies. However, given that the models were developed using a single state's data, the HSM recommends calibration of the prediction models using data from the jurisdiction where they will be applied. This calibration process has been conducted in several states and many questions have been raised as a result. This paper is intended to investigate different definitions and criteria for the calibration factors, and provide recommendations for practitioners on which definition to use. In addition to the calibration factors in the HSM and previously published definitions, two other calibration factor equations are proposed and compared using multiple goodness of fit measures. Whereas each definition may outperform others in certain measures, in this study, it is recommended to use the definition that maximizes the likelihood between predicted and observed crashes. The idea is to follow the same concept in both state-specific safety performance functions development and calibration process, which is maximizing the likelihood of predicted and observed crashes.

On average in the last 20 years in the United States, more than 39,000 persons have lost their lives due to highway motor vehicle traffic crashes each year (*1*). Total U.S. highway fatalities have been decreasing by an average of about 500 fatalities per year (*1*). Among numerous factors that contributed to this reduction, crash frequency prediction had a significant role with its numerous applications in highway design and network screening.

The most commonly accepted crash frequency prediction models are documented in the *Highway Safety Manual* (HSM), which was initially published by AASHTO in 2010. In Part C of the HSM, models are documented to predict the crash frequency for different facility types (*2*). These predictive models consist of three main components: safety performance functions (SPFs), crash modification factors (CMFs), and calibration factors (CFs). The SPFs are the regression models used to predict the crash frequency for base conditions, the CMFs are adjustment factors/functions to account for the site's specific design and development conditions, and the CFs account for the factors that are not included in the model but are expected to be significant such as weather, crash reporting systems, animal populations, and so forth. The predicted crashes (N_p) in the HSM are defined as the following:

$$N_p = N_{spf} \times (CMF_1 \times CMF_2 \times \ldots) \times CF \tag{1}$$

where

N_p = predicted average crash frequency,

N_{spf} = predicted average crash frequency for base condition,

CMF_i = crash modification factor,

CF = calibration factor.

Before using the prediction models in the HSM, it is recommended to calibrate these models using data from the jurisdiction where they will be applied (*2*). The CF is defined as the summation of the observed crashes (N_o) divided by the summation of unadjusted predicted crashes (N_u) over the selected samples (*2*). This definition is noted hereafter as CF_{HSM} and is calculated as follows:

$$CF_{HSM} = \frac{\sum_{i=1}^{N} N_{oi}}{\sum_{i=1}^{N} N_{ui}} \tag{2}$$

where

[1]Civil Engineering, Clemson University, Clemson, SC
[2]Applied Statistics, Clemson University, Clemson, SC

Corresponding Author:
Address correspondence to Mahdi Rajabi: mrajabi@clemson.edu

$N_u = N_{spf} \times (CMF_1 \times CMF_2 \times \ldots)$,
N_o = number of Observed crashes,
N_u = number of Unadjusted predicted crashes,
N = sample size.

Using CF_{HSM}, the mean of the predicted ($\bar{N}_p$) and the mean of observed crashes ($\bar{N}_o$), over the calibration sample, will be equal. In other words, CF_{HSM} minimizes the absolute summation of error between the observed and the predicted crashes:

$$\left|\sum_{i=1}^{N}(N_{oi} - N_{pi})\right| = \left|\sum_{i=1}^{N}(N_{oi} - CF_{HSM} \times N_{ui})\right| = \left|\sum_{i=1}^{N}\left(N_{oi} - \frac{\sum_{j=1}^{N} N_{oj}}{\sum_{j=1}^{N} N_{uj}} \times N_{ui}\right)\right| = 0 \quad (3)$$

In a study by Mehta and Lou (*3*), a new CF is defined as a special case of the SPF development. In this study, the CF is defined as the intercept of a negative binomial (NB) regression, in which N_o is the dependent variable, N_u is the independent variable, and CF is the only parameter to estimate. This CF is noted hereafter as CF_{Mehta} and can be obtained from the following:

$$N_o = e^{\ln(CF_{Mehta}) + \ln(N_u)} \quad (4)$$

In the aforementioned study, data for rural two-lane-undivided (R2U) and rural four-lane-divided (R4D) roadways, from Alabama, were utilized to analyze the performance of CF_{Mehta} (*3*). The comparison was performed using three goodness of fit (GoF) measures introduced by Washington et al. including mean absolute deviation (MAD), mean prediction bias, and mean squared prediction error (these three measures are defined later in this paper) plus log likelihood (LL) and Akaike's information criterion (AIC) (*4*). Based on the results, CF_{Mehta} outperformed CF_{HSM} in LL and AIC, whereas CF_{HSM} outperformed CF_{Mehta} in three GoF measures. In the study by Mehta and Lou, state-specific SPFs were developed, but the HSM-calibrated SPF predictions were not compared with state-specific SPF predictions (*3*).

In another study by Srinivasan et al. using rural two-lane-undivided (R2U) highways of Arizona, categorical CFs and calibration functions are developed (*5*). For categorical CFs, in addition to an overall CF, CFs were estimated by highway functional class, annual average daily traffic (AADT) categories (e.g., AADT < 2,500; 2,500 < AADT < 5,000; and AADT > 5,000), segment length, and so on. The performance of categorical CFs was compared with the overall CF using cumulative residuals (CURE) plots (*6*), in which the categorical CFs were not found to achieve significant improvement. As a result, the calibration functions were investigated and six different forms of calibration functions were developed, including 2 to 4 parameters, estimated by NB regression. According to these results, the calibration functions were preferred based on CURE plots compared with the overall CF or the categorical CF_{HSM}.

This paper aims to provide an alternative to the CF_{HSM} and analyze the performance of the proposed definition using several statistical methods. Furthermore, because state-specific SPFs provide more reliable predictions compared with calibrated HSM SPFs (*2*), they are used as a reference for comparing the CFs. This paper attempts to define a CF that maximizes the likelihood of the predicted crashes, assuming that crashes have a NB distribution. This is the same concept that is used in the SPF development process where coefficients of regression are obtained using maximum likelihood estimation (MLE). In this study, a CF is defined using the MLE method. For this purpose, the likelihood function is derived based on the predicted and observed crashes, assuming a NB distribution. Then, the CF that maximizes the likelihood function is defined as the alternative CF. This CF is called CF_{MLE} and can be obtained using the following equation:

$$N = \sum_{i=1}^{N} \frac{N_{oi} + \varphi}{CF_{MLE}N_{ui} + \varphi} = \sum_{i=1}^{N} \frac{N_{oi} + \varphi}{N_{pi} + \varphi}$$
$$\mathrm{var}(N_o) = \bar{N}_o + \frac{\bar{N}_o^{\,2}}{\varphi} \quad (5)$$

where

φ = inverse dispersion parameter of observed crashes
$\bar{N}_o$, $\mathrm{var}(N_o)$ = mean and variance of observed crashes respectively
N = sample size.

The above equation does not provide an explicit definition for CF_{MLE}, however, this can be easily solved for CF_{MLE} using numerical methods. The maximum likelihood estimate, obtained from CF_{MLE}, is expected to provide the closest likelihood to the state-specific SPF likelihood. The likelihood obtained from the calibrated HSM predictions is expected to be lower than (or at most equal to) the likelihood obtained from state-specific SPFs, because the CFs, in general, just scale the HSM SPFs, whereas the state-specific SPFs are more flexible (and complex) and can change the functional form as well.

Another alternative CF is defined using the least squares estimation (LSE) method. This CF minimizes the summation of squared deviation between the observed crashes and the predicted crashes, and can be obtained using the following equation:

$$CF_{LSE} = \frac{\sum_{i=1}^{N}(N_{oi} \times N_{ui})}{\sum_{i=1}^{N} N_{ui}^{2}} \quad (6)$$

Data from the South Carolina calibration and SPF development project (*7*) is used to examine the four CF alternative definitions including: HSM, MLE, LSE, and Mehta. To compare different CFs, in addition to the three aforementioned GoF measures by Washington et al. (*4*), other criteria are considered in this study including the absolute deviance between the average state fitted crashes ($\bar{N}_f$) and the average calibrated predicted crashes ($|\bar{N}_f - \bar{N}_p|$), and CURE plots. State fitted crashes ($\bar{N}_f$) are the predicted values of the state-specific SPFs. After a brief literature review, new definitions of the CFs are described in detail, followed by the comparison results, and conclusions.

Previous Research

The idea of developing the HSM as a document to quantify the effects of highway design on safety grew out of a conference session in the 78th Annual Meeting of the Transportation Research Board (TRB) in January 1999 (*8*). As part of the efforts to develop the HSM, an extensive HSM user survey was conducted by Harwood et al., which found crash frequency to be the top priority output variable of the HSM (*8*). The HSM was later published in 2010 providing prediction models for crash frequency. These prediction models are provided in Part C of the HSM and cover eight types of roadways and 10 types of intersections (*2*).

The HSM recommends calibrating the prediction models or developing jurisdiction-specific models, for which data and expertise are available. To date, several states have conducted HSM model calibration (or state-specific model development, or both HSM model calibration and state-specific model development) studies including but not limited to Alabama, South Carolina, Oregon, North Carolina, Florida, Missouri, Maryland, Illinois, and Louisiana (*3, 7, 9–16*). In addition to the calibration projects, there have been only two studies focusing on the CF definition, which will be discussed further in this paper.

In a study by Mehta and Lou, which developed calibration factors and state-specific SPFs for two roadway types in the state of Alabama (R2U and R4D), a new CF was defined as a special case of SPF development as the following (also mentioned in Equation 4) (*3*).

$$N_o = e^{\ln(CF_{\text{Mehta}}) + \ln(N_u)} \tag{7}$$

One should note that CF_{Mehta} is defined as the intercept of the NB regression, whereas N_u is the offset variable. It is noteworthy that by disregarding the assumption of the NB regression the above equation can be simplified as the following:

$$N_o = e^{\ln(CF_{\text{Mehta}}) + \ln(N_u)} \overset{\text{Relaxing the NB regession assumption}}{\Rightarrow} N_o/N_u = CF_{\text{Mehta}} \Rightarrow CF_{\text{Mehta}} = \frac{\sum_{i=1}^{N} N_{oi}/N_{ui}}{N} \tag{8}$$

where CF_{Mehta} = the CF_{Mehta} obtained by relaxing the assumption of NB regression.

In other words, by relaxing the assumption of NB regression, CF_{Mehta} calculates the CF for each individual site, using the HSM definition, and then takes the average.

In another study by Srinivasan et al., two different approaches were examined for the CF definitions using 196 Arizona R2U sites with a total of 187.5 mi (*5*). In the first approach, the overall CF, using the HSM definition, was computed for the entire sample and then the subsets of the sample were considered for categorical CFs including subsets based on AADT, length, horizontal alignment, curve radius, and so forth. It was shown that even though the overall CF might be very close to 1, for different categories the CF may change significantly. The performance of categorical CFs was examined using CURE plots, which indicated that categorical CFs would not be preferred.

In the second approach by Srinivasan et al., six different types of calibration functions were investigated (*5*). For example, the first type of calibration function is defined as the following:

$$N_p = a \times N_u^{\,b} \tag{9}$$

where $a.b$ = regression parameters.

In this calibration function, all the components in N_u are assumed to have the same power, the other five calibration functions are defined by relaxing this assumption, and as a result have more parameters and are more flexible. The parameters of the calibration functions are also calculated by NB regression. The calibration functions provide more accurate predictions compared with CFs, and are comparable with state-specific SPFs (rather than CFs) but the parameter estimation for the calibration functions may require similar statistical expertise, as does state-specific SPF development.

Methodology

In this part, first the LL function of number of observed crashes, given a set of predicted crashes and assuming a NB distribution, is derived and then CF_{MLE} is calculated by maximizing the obtained LL function. Next, using the LSE method, CF_{LSE} is derived. Afterwards, the data from the South Carolina calibration and SPF development project (*7*) is utilized to assess the four different

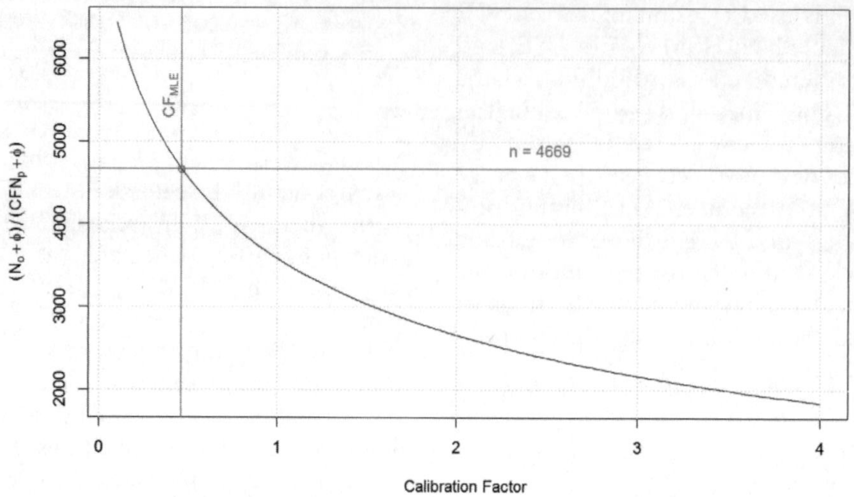

Figure 1. CF_{MLE} calculation for R3ST intersections.

definitions for the CFs (CF_{HSM}, CF_{MLE}, CF_{LSE}, and CF_{Mehta}).

Calibration Factor by MLE Method

The probability mass function of the NB distribution, assuming y as the number of success, p as the probability of success and r as the number of failures, using gamma function (Γ) can be written as the following:

$$f(y;r.p) = \frac{\Gamma(y+r)}{\Gamma(y+1)\Gamma(r)}(p)^y(1-p)^r \tag{10}$$

In crash analysis terms and assuming success equals crash, y can be replaced by number of observed crashes (N_{oi}). Given that the mean and the variance of the NB distribution are $\mu = {pr}/{1-p}$ and $\sigma^2 = {pr}/{(1-p)^2}$ respectively, and the inverse dispersion parameter (φ) is defined as $\sigma^2 = \mu + {\mu^2}/{\varphi}$, r and p can be replaced by φ and ${\mu}/{(\mu+\varphi)}$ respectively. Also μ can be replaced by $N_p = CF \times N_u$ because N_p is the expected value of N_o. As a result, the above probability mass function can be written as follows (please note that φ is a function of N_o):

$$f(N_{oi};N_{pi}) = \frac{\Gamma(N_{oi}+\varphi)}{\Gamma(N_{oi}+1)\Gamma(\varphi)}\left(\frac{N_{pi}}{N_{pi}+\varphi}\right)^{N_o}\left(\frac{\varphi}{N_{pi}+\varphi}\right)^{\varphi} \tag{11}$$

The likelihood function can be derived as the multiplication of the probability of all observed crashes as follows:

$$l(N_o,N_p) = \prod_{i=1}^{N}\left(\frac{\Gamma(N_{oi}+\varphi)}{\Gamma(N_{oi}+1)\Gamma(\varphi)}\left(\frac{N_{pi}}{N_{pi}+\varphi}\right)^{N_{oi}}\left(\frac{\varphi}{N_{pi}+\varphi}\right)^{\varphi}\right) \tag{12}$$

The natural logarithm of the above equation is used as the desired LL function. Then, CF_{MLE}, is obtained by replacing N_p by $CF \times N_u$ and solving $\frac{\partial l}{\partial CF} = 0$ as follows (also presented in Equation 5):

$$N = \sum_{i=1}^{N}\frac{N_{oi}+\varphi}{CF_{\mathrm{MLE}}N_{ui}+\varphi} = \sum_{i=1}^{N}\frac{N_{oi}+\varphi}{N_{pi}+\varphi} \tag{13}$$

This formula needs to be solved numerically for CF_{MLE}. To better understand the behavior of the equation above, in Figure 1, the summation term (right side of the equation) is plotted against a range of the CFs, using data from rural three-leg-stop-controlled (R3ST) intersections in South Carolina. A horizontal line representing the sample size (left side of the equation) is also shown and the intersection point is identified as CF_{MLE}.

As shown in Figure 1, the smooth and continuous behavior of the summation term allows Equation 13 to be easily solved by several numerical methods (e.g., bisection method) or common computer programs (e.g., Excel).

Calibration Factor by LSE Method

The LSE method is known as the standard method to find the regression parameters, especially in linear models. In this section, this method is used to calculate the

Table 1. Calibration Factors For Roadway Types

	Type	R2U	R4D	R4U	U2U	U3T	U4U	U4D	U5T
Roadways	N	1,234	339	142	447	24	234	235	454
	Mileage	751.5	108.9	37.4	139.4	6.2	52.1	58	106.6
	$\bar{N}_o$	0.2428	0.4980	0.1198	0.3913	1.1233	0.7880	0.9119	1.5379
	$\bar{N}_u$	0.2451	0.8121	0.3901	0.2361	0.7633	1.0520	1.1008	2.0024
	$\bar{N}_f$	0.2404	0.5204	0.1200	0.4001	1.0581	0.8227	0.9179	1.5188
	CF_{HSM}	0.9907	0.6133	0.3072	1.6574	1.4717	0.7490	0.8284	0.7680
	CF_{MLE}	1.1166	0.6832	0.3002	1.9396	1.1125	0.8205	0.8678	0.7830
	CF_{LSE}	0.8507	0.4798	0.3277	1.3509	1.8928	0.6808	0.7695	0.7556
	CF_{Mehta}	1.0386	0.6707	0.3055	1.7533	1.3017	0.7776	0.8768	0.7749

Table 2. Calibration Factors For Intersection Types

	Type	R3ST	R4ST	R4SG	RM3ST	RM4ST	RM4SG	UM3ST	UM4ST	UM3SG	UM4SG
Intersections	N	4,669	1,858	65	429	191	53	3,759	1,994	201	360
	$\bar{N}_o$	0.1296	0.2826	1.3505	0.4258	0.2218	3.4000	0.3810	0.5515	4.1973	6.1970
	$\bar{N}_u$	0.3218	0.5960	2.9611	0.7690	0.8586	8.5310	0.3179	0.5746	2.1020	2.5318
	$\bar{N}_f$	0.1304	0.2918	1.3694	0.4091	0.2232	3.4184	0.3702	0.5273	4.1334	6.0909
	CF_{HSM}	0.4026	0.4741	0.4561	0.5537	0.2583	0.3985	1.1985	0.9598	1.9968	2.4477
	CF_{MLE}	0.4109	0.4282	0.4659	0.4679	0.2382	0.3889	1.1693	0.7845	1.8889	2.3143
	CF_{LSE}	0.3824	0.5164	0.4333	0.6877	0.2971	0.4085	1.0751	1.1422	2.1594	2.5934
	CF_{Mehta}	0.4039	0.4689	0.4691	0.5413	0.2563	0.3865	1.2101	0.9377	1.8905	2.3302

CF, called CF_{LSE}. For this purpose, the first derivative of the summation of the squared error between observed and predicted crashes with respect to the CF is obtained and solved for the CF as follows:

$$SSE = \sum_{i=1}^{N} (N_{oi} - CF \times N_{ui})^2$$

$$\frac{\partial SSE}{\partial CF} = 0 \Rightarrow CF_{LSE} = \frac{\sum_{i=1}^{N}(N_{oi} \times N_{ui})}{\sum_{i=1}^{N}(N_u{}^2)} \quad (14)$$

Summary of the Data

The aforementioned CFs were evaluated using data from the South Carolina HSM calibration project (*7*). For this purpose, the facility types in the HSM Part C, are abbreviated using three main components:

1. Development level: rural or urban (R or U).
2. Number of lanes or legs: for roadway segments two, three, four, five, or M (multiple lanes); or for intersections, three or four indicating the number of approach legs.
3. Median barrier/traffic control: for roadway segments undivided (U), divided (D), or two-way left-turn lane (T); for intersections, either stop (ST) or signal controlled (SG).

Tables 1 and 2 provide the values of the aforementioned CF definitions for roadways and intersection facility types with a summary of average crashes for each site.

Comparing Calibration Factors

In this section, the six methods used for CF comparison are described and include: the LL function, absolute deviance between the mean of state fitted crashes and the mean of predicted crashes ($|\bar{N}_f - \bar{N}_p| = |\bar{N}_f - CF \times \bar{N}_u|$), MAD, sum of squared error (SSE), coefficient of variation (CV), and CURE plots. After the description of each method, the GoF measure is calculated for each alternative CF definition and the results are briefly discussed.

Comparing by Log Likelihood. First, the CFs were compared with the LL function introduced in Equation 12. To provide an example of how the likelihood function changes for different values of CFs, the LL function was plotted against CF for the R2U roadways of South Carolina (Figure 2). Additionally, the LL of the state-specific SPF plus all five CFs are shown in the plot.

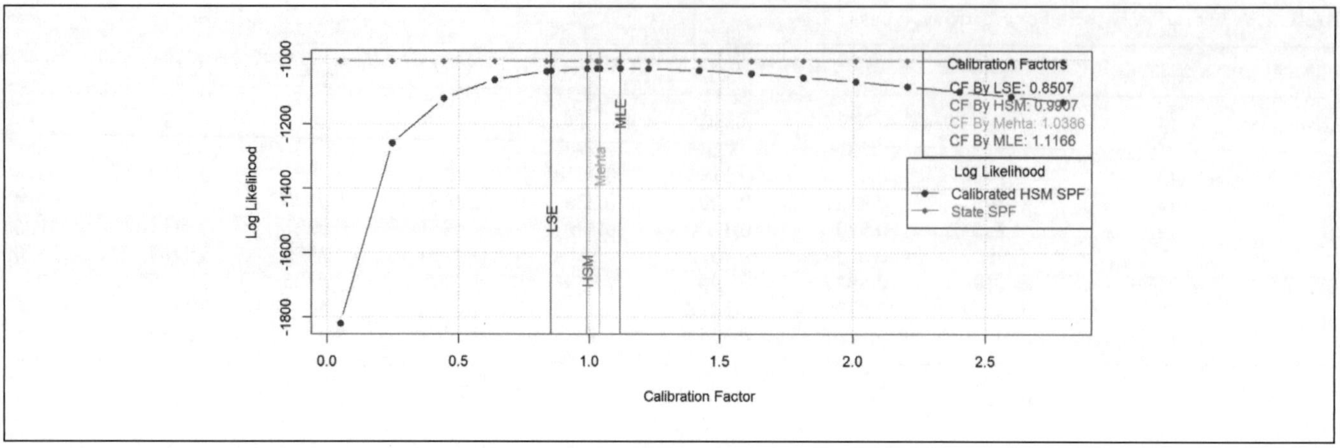

Figure 2. Log-likelihood function for R2U roadways.

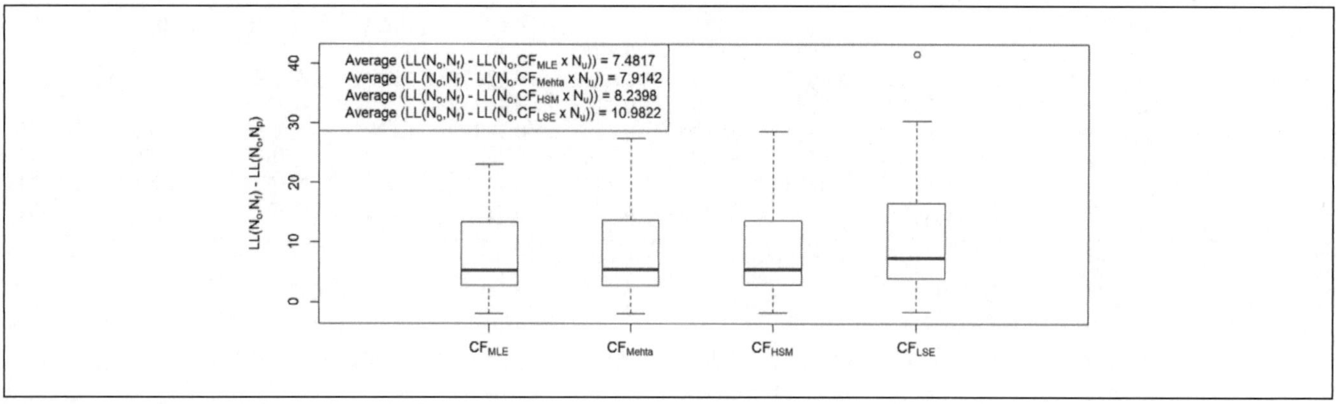

Figure 3. CFs comparison based on the deviance between $ll(N_o, N_f)$ and $ll(N_o, N_p)$.

Figure 2 shows how CF_{MLE} maximizes the LL function as described in Equation 13. Moreover, it shows that the calibrated HSM SPFs, even if calibrated with CF_{MLE}, cannot provide a greater likelihood than state-specific SPFs. The state-specific SPF's LL remains as a horizontal line against CF, because it is not related to the CF as it is calculated based on state fitted crashes and observed crashes. The same analysis was conducted on all 18 facility types and the difference between the likelihood of the state fitted values and predicted crashes, $ll(N_o.N_f) - ll(N_o.CF \times N_u)$, was obtained. The results were sorted based on the average deviance, and corresponding boxplots are shown in Figure 3.

Figure 3 shows that on average, CF_{MLE} provides the closest likelihood estimate compared with the state-specific SPF likelihood among all CF definitions. It should be mentioned that the calibration functions by Srinivasan and Bauer were expected to provide better likelihood estimates than any aforementioned CF, because they implement more than one parameter and are not shown in the figure above (*17*). However, the average deviance between the LL estimates of the state fitted crashes and the Type 1 calibration function predictions were calculated and were 3.2 compared with 7.4, which corresponds to CF_{MLE}.

Comparing by Absolute Deviance. Next, CFs were compared using absolute deviance between the mean of state fitted crashes and the mean of predicted crashes ($|\bar{N}_f - \bar{N}_p| = |\bar{N}_f - CF \times \bar{N}_u|$). The deviance for CFs was calculated for all 18 facility types and the results are presented in Figure 4. The CFs were sorted based on the mean of deviance from lowest to highest.

Figure 4 indicates that CF_{HSM} provides the closest mean of predicted crashes to the mean of state fitted crashes over the sample data. This can be explained by considering that the mean of state fitted values is very close to the mean of observed crashes.

Comparing by Mean Absolute Deviance. In addition, MAD was considered for comparing the CFs. This GoF measure was proposed by Washington et al. for validating the accident models of intersections and has been used

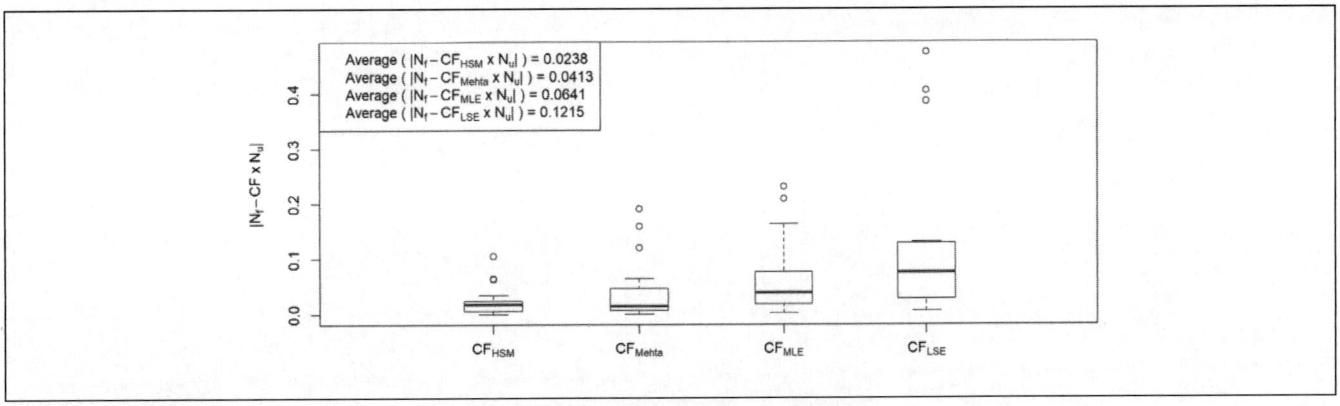

Figure 4. Calibration factors comparison based on the deviance between $\bar{N}_f$ and $\bar{N}_p$.

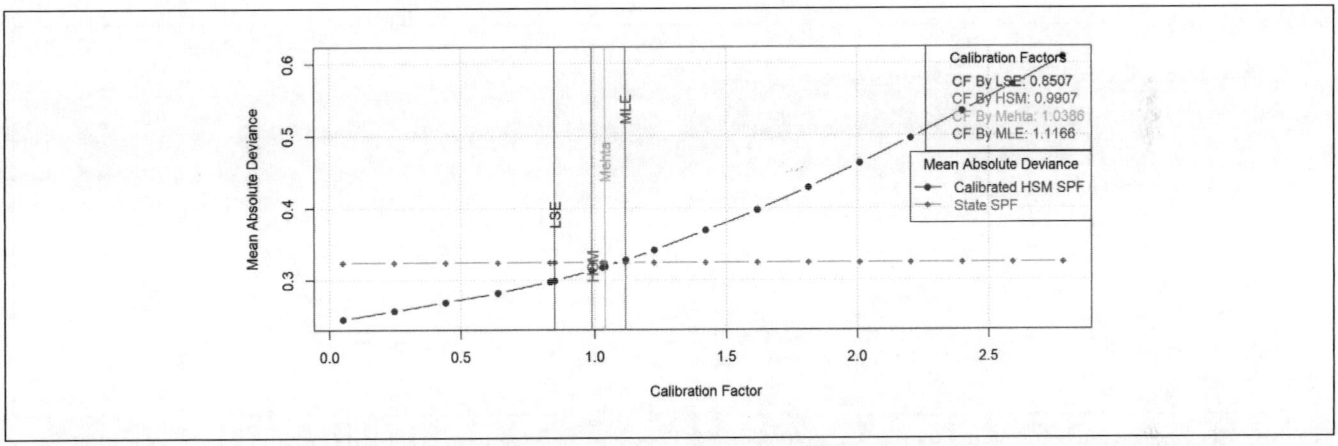

Figure 5. The mean absolute deviance between observed and predicted crashes for R2U roadways.

by other studies to measure GoF (*3*, *4*). MAD is defined as follows:

$$MAD = \frac{\sum_{i=1}^{N} |N_{oi} - N_{p_i}|}{N} \quad (15)$$

For instance, MAD was calculated for R2U roadways and is plotted against CF in Figure 5. MAD based on the state fitted crashes is also shown in Figure 5, as well as the four definitions of CFs.

The minimum of MAD in R2U roadways did not occur in the common range of the CFs (˜0 to 5). MAD was calculated for all other facility types and it was observed that in many cases it either did not have any minimum in the common range of the CFs or the minimum was very close to zero. As a result, the authors of this paper found MAD to be an inappropriate measure for this purpose and decided not to compare the CFs using MAD.

Comparing by Sum of Squared Error. The CFs were then compared using SSE between observed and predicted crashes. SSE can be obtained using the following equation:

$$SSE = \sum_{i=1}^{N} (N_{oi} - N_{p_i})^2 \quad (16)$$

SSE was calculated for R2U roadways of South Carolina and is plotted against the CF in Figure 6.

As shown in Figure 6, CF_{LSE} minimizes SSE, which was expected based on its definition. Unlike LL, SSE obtained from HSM-calibrated SPFs might be lower than SSE obtained from the state-specific SPFs because in the regression process, the coefficients are calculated to maximize the LL. Nevertheless, the SSE calculation was repeated over all 18 facility types and the results are shown in Figure 7. As expected, on average CF_{LSE} provided the lowest SSE compared with the other CF definitions.

Comparing by Coefficient of Variation. The CFs were also compared with their CV, which is defined as the

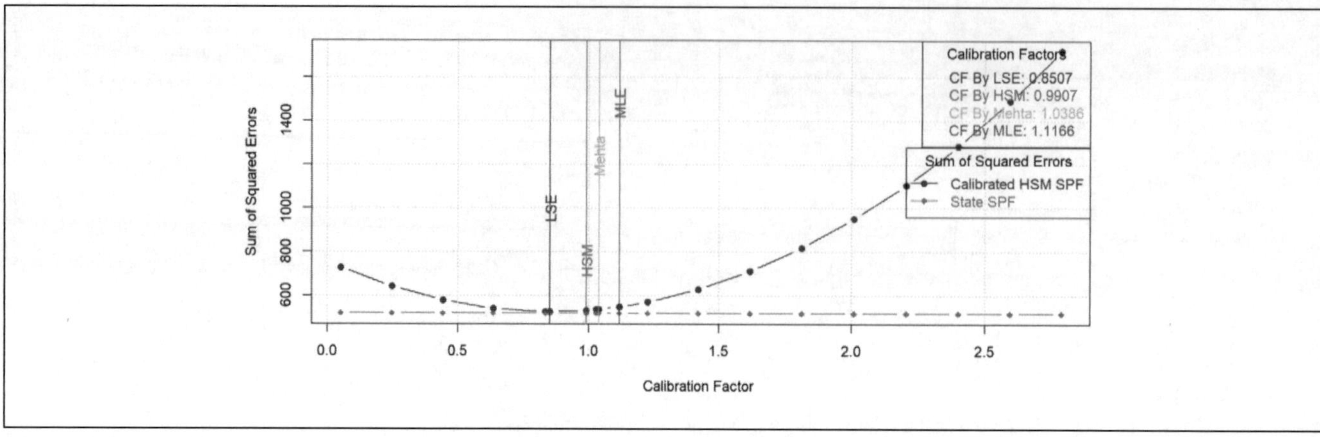

Figure 6. The sum of squared error (SSE) between observed and predicted crashes for R2U roadways.

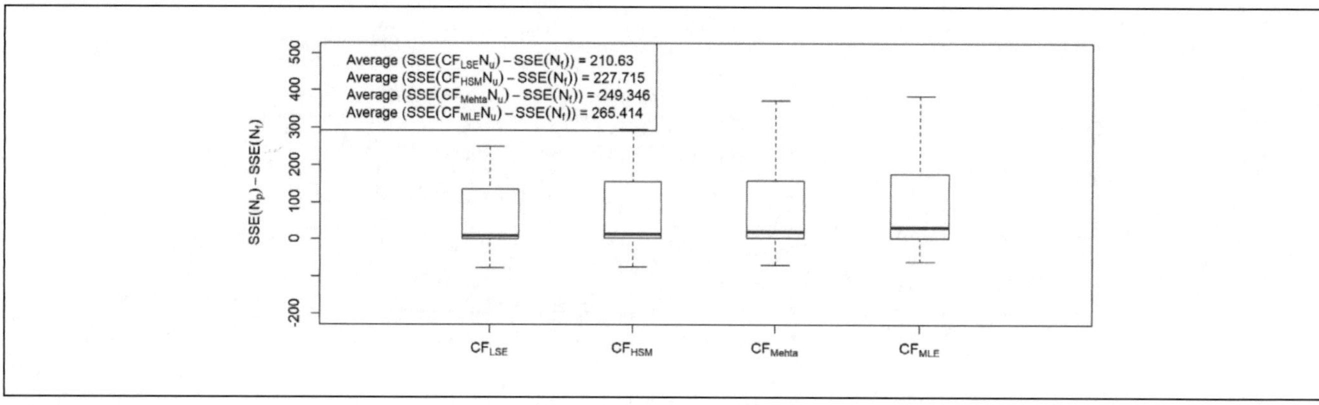

Figure 7. CFs comparison based on the sums of squared error (SSE) between observed and predicted crashes.

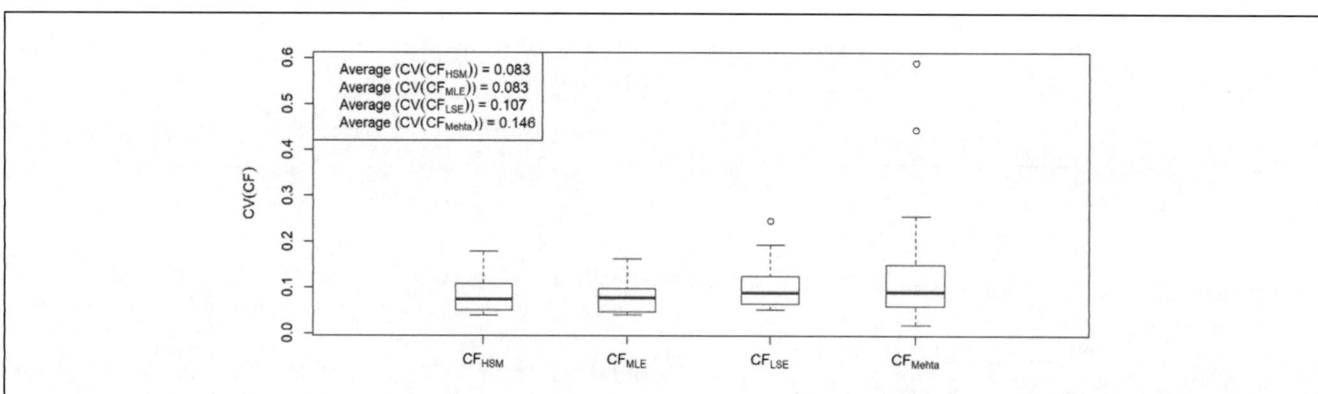

Figure 8. CFs comparison based on coefficient of variation.

standard deviation divided by the mean ($CV = {}^{s}/_{\mu}$). Figure 8 shows the average of the CV of the CFs over all 18 facility types. In this comparison, CF_{HSM} and CF_{MLE} were almost equal in terms of the CV.

Comparing by Cumulative Residual Plots. Lastly, the CFs were compared using CURE plots; however, CURE plots are mostly used to assess the functional form of the model and by changing the value of the CF, the functional form of the model does not change. Thus, the CFs only shift the CURE plots up and down. The CURE plots of state fitted crashes (N_f) were compared with the CURE plots of predicted crashes (N_p) using each CF definition. Although the shape of the CURE plot is a

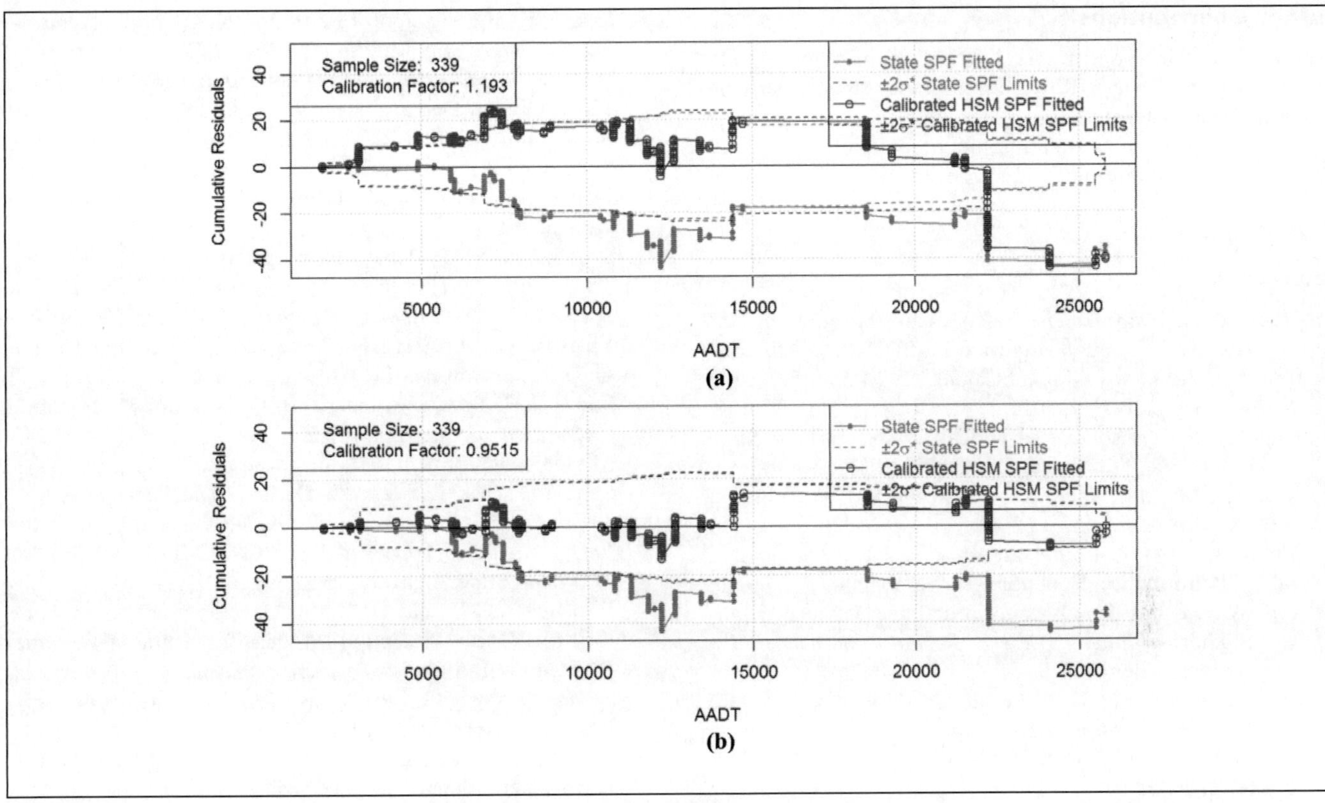

Figure 9. CURE plot of R4D roadways: (*a*) HSM SPF calibrated by $\boldsymbol{CF_{MLE}}$; (*b*) HSM SPF calibrated by Srinivasan et al. (*5*) calibration function Type 1.
Note: AADT = annual average daily traffic.

function of how well the model fits the data, the last point (or the landing point) of the CURE plot (the value of cumulative residuals corresponding to the maximum value of the predictor) is a function of the mean of the predictions. As a result, if the mean of HSM-calibrated SPFs predictions (N_p) is lower than the mean of state-specific SPF predictions (N_f), the CURE plot of HSM-calibrated SPF will land higher (assuming positive residuals at the top of the vertical axes) than the CURE plot of state-specific SPF and vice versa. The CURE plots for R4D roadways are shown for CF_{MLE} and calibration function Type 1 by Srinivasan et al. in Figure 9 (*5*).

As is shown in Figure 9, although the CFs only shift the CURE plots up and down, the calibration function changes the shape of the CURE plot.

Summary and Conclusion

Based on the results of this study, to provide the most reliable predictions for crash frequency, state-specific SPFs should be developed. Agencies may need enough data and analysts with the statistical expertise to develop state-specific SPFs (*17*, *18*). The next priority will be developing calibration functions which could be developed with lower data requirements but well-trained analysts (*5*). If agencies decide to have a calibration factor, the results of this study show that using CF_{MLE} may lead to predictions that have higher likelihood compared with CF_{HSM} and is recommended because by definition, CF_{MLE} will always provide the highest likelihood compared with other CFs.

It was shown in this study that depending on the measure considered, different definitions of CFs may outperform others, for instance, CF_{MLE} in likelihood, CF_{HSM} in absolute summation of error, and CF_{LSE} in summation of squared error. To find the right measure for developing the CF, the idea in this study was to consider the same likelihood function that is used for developing state-specific SPFs. Thus, by definition, the crash frequency predictions based on CF_{MLE} will provide the highest likelihood compared with other CF definitions.

Acknowledgment

The authors thank the South Carolina Department of Transportation for funding the South Carolina HSM calibration project and providing the data for this study.

Author Contributions

The authors confirm contribution to the paper as follows: study conception and design: MR; performing analysis and deriving equations: MR; interpretation of results and supervising the process: JO, PG; draft manuscript preparation: MR. All authors reviewed the results and approved the final version of the manuscript.

References

1. National Highway Traffic Safety Administration. *FARS Encyclopedia.* National Highway Traffic Safety Administration. http://www-fars.nhtsa.dot.gov/Main/index.aspx. Accessed September 19, 2016.
2. *Highway Safety Manual.* American Association of State Highways and Transportation Officials, Transportation Research Board of the National Academies, Washington D.C., 2010.
3. Mehta, G., and Y. Lou. Calibration and Development of Safety Performance Functions for Alabama: Two-Lane, Two-Way Rural Roads and Four-Lane Divided Highways. *Transportation Research Record: Journal of the Transportation Research Board*, 2013. 2398: 75–82.
4. Washington, S. P., B. N. Persaud, C. Lyon, and J. Oh. *Validation of Accident Models for Intersections.* FHWA-RD-03-037. Federal Highway Administration, Washington, D.C., 2005.
5. Srinivasan, R., M. Colety, G. Bahar, B. Crowther, and M. Farmen. Estimation of Calibration Functions for Predicting Crashes on Rural Two-Lane Roads in Arizona. *Transportation Research Record: Journal of the Transportation Research Board*, 2016. 2583: 17–24.
6. Hauer, E., and J. Bamfo. Two Tools for Finding What Function Links the Dependent Variable to the Explanatory Variables. Presented at ICTCT Workshop in Lund, Sweden, 1997.
7. Ogle, J. H., and M. Rajabi. *Development of South Carolina Database and Calibration Factors For The Highway Safety Manual.* South Carolina Department of Transportation, 2017.
8. Harwood, D. W., K. M. Bauer, K. R. Richard, D. K. Gilmore, J. L. Graham, I. B. Potts, D. J. Torbic, and E. Hauer. *NCHRP Report 129: Methodology to Predict the Safety Performance of Urban and Suburban Arterials.* TRB, National Research Council, Washington, D.C., 2007.
9. Dixon, K., C. Monsere, F. Xie, and K. Gladhill. *Calibrating the Future Highway Safety Manual Predictive Methods for Oregon State Highways.* Publication FHWA-OR-RD-12-07. Oregon Department of Transportation, 2012, p. 98.
10. Srinivasan, R., and D. L. Carter. *Development of Safety Performance Functions for North Carolina.* North Carolina Department of Transportation, Research and Analysis Group, 2011.
11. Srinivasan, S., P. Haas, N. S. Dhakar, R. Hormel, D. Torbic, and D. Harwood. *Development and Calibration of Highway Safety Manual Equations for Florida Conditions.* Final Report, Project BDK77 977-06. Florida Department of Transportation, 2011.
12. Sun, C., H. Brown, P. Edara, B. Claros, and K. A. Nam. *Calibration of the Highway Safety Manual for Missouri.* Missouri Department of Transportation, 2013.
13. Shin, H., Y. J. Lee, and S. Dadvar. *The Development of Local Calibration Factors for Implementing the Highway Safety Manual in Maryland.* Report No. MD-14-SP209B4J. Maryland State Highway Administration, Hanover, 2014.
14. Williamson, M., and H. Zhou. Develop Calibration Factors for Crash Prediction Models for Rural Two-Lane Roadways in Illinois. *Procedia – Social and Behavioral Sciences*, Vol. 43, 2012, pp. 330–338. https://doi.org/10.1016/j.sbspro.2012.04.106.
15. Zhao, J. Calibrating the Highway Safety Manual Safety Performance Function with Crash Data in Illinois, Illinois Department of Transportation, 2013.
16. Sun, X., D. Magri, H. H. Shirazi, S. Gillella, and L. Li. Application of Highway Safety Manual: Louisiana Experience with Rural Multilane Highways. Presented at 90th Annual Meeting of the Transportation Research Board, Washington, D.C., 2011.
17. Srinivasan, R., and K. Bauer. *Safety Performance Function Development Guide: Developing Jurisdiction-Specific SPFs.* Publication FHWA-SA-14-005. Federal Highway Administration Office of Safety, Washington, D.C., 2013.
18. Srinivasan, R., D. Carter, and K. Bauer. *Safety Performance Function Decision Guide: SPF Calibration vs SPF Development.* Publication FHWA-SA-14-004. Federal Highway Administration Office of Safety, Washington, D.C., 2013.

The Standing Committee on Highway Safety Performance (ANB25) peer-reviewed this paper (18-02091).

TRR
JOURNAL OF THE TRANSPORTATION RESEARCH BOARD

Article

Transportation Research Record
2018, Vol. 2672(30) 109–119
© National Academy of Sciences:
Transportation Research Board 2018
Article reuse guidelines:
sagepub.com/journals-permissions
DOI: 10.1177/0361198118794052
journals.sagepub.com/home/trr
SAGE

A Comparative Analysis on Performance of Severe Crash Prediction Methods

Raul E. Avelar[1], **Karen Dixon**[1], **and Sruthi Ashraf**[1]

Abstract

The objective of this paper is to compare the performance and tradeoffs between two alternative analysis methods for developing crash prediction models for severe crashes: a direct estimation of severe crashes using frequency models, and an indirect but popular approach of combining frequency of total crashes models and some form of severity distribution functions (SDFs). The researchers conducted a comprehensive comparison of these modeling methods to illustrate the strengths and weaknesses of each alternative, and to inform future research that intends to develop such models. An examination of the theoretical characteristics of the modeling approach is presented and discussed. The performance of the two modeling alternatives is compared using two different datasets. The results of those comparisons showed very similar performances by both techniques. Finally, a sensitivity analysis is presented to explore how the performance of these techniques vary by degree of dispersion and observed correlation levels of total and severe injury crashes (KAB; injury scale in which K = fatal [killed], A = incapacitating injury, B = nonincapacitating injury) with potential explanatory variables. The results from these analyses tended to favor the use of SDFs in combination with total crashes safety performance functions (SPFs), as the prediction tended to show reduced dispersion under most conditions. However, performance of the KAB SPF model outperformed the combination of SDF and SPF for total crashes when KAB and non-KAB crashes had a common predictor but with effects in opposite directions.

The objective of this paper is to compare the performance and tradeoffs among two alternative analysis methods for developing crash prediction models for severe crashes: a direct estimation of severe crashes using frequency models, and a more indirect but popular approach of combining frequency of total crashes models and some form of severity distribution functions (SDFs). The researchers conducted a comprehensive comparison of model methods to illustrate strengths and weaknesses of each alternative and to inform future research that intends to develop such models.

Literature Review

This section presents a summary of literature on previous studies that analyzed the severity of crashes. In addition to reviewing works about methodological differences in analyzing severe crashes, the focus of this review is to document two types of studies: (a) those that developed models on proportions of severe crashes, and (b) those that treated frequency of crashes of different severity levels in separate models.

Methodological Comparisons

SDFs are used to estimate the crash severity as a function of geometric and traffic characteristics (*1*) and must be calibrated when applied to a new jurisdiction. SDFs vary differently with traffic volumes as well as design decisions depending on the crash type of interest (*2*). As the SDF model considers all severity levels together, the SDFs allow predicting the shift in crashes among levels due to roadway condition changes (*3*). Lord et al. estimated the required sample size and the factors that influence the selection of the sample size for the SDFs calibration or recalibration (*4*). The sample size increased with increasing coefficient of variation and confidence level.

Crash severity analysis models can be generally classified as either nominal or ordinal. The three most common nominal models used are multinomial logit (MNL), nested logit, and mixed logit models whereas the major

[1]Texas A&M Transportation Institute, College Station, TX

Corresponding Author:
Address correspondence to Raul E. Avelar: r-avelar@tamu.edu

ordinal models are ordered logit, ordered probit, and ordered mixed logit models (*5*). In a comparison of sample size requirements by Lord and Ye among MNL ordered probit and mixed logit models, mixed logit was found to require the largest sample size and ordered probit model the smallest (*6*).

Park and Lord modeled correlated crash frequencies at varying severity levels using a multivariate Poisson-lognormal (MVPLN) specification (*7*). Ma et al. evaluated this approach within the Bayesian framework (*8*). Not surprisingly, both analyses showed that statistically significant correlations exist between crash counts for different levels of injury severity. Xie et al. applied a Bayesian ordered probit model to analyze driver's injury severity and compared it with the results obtained from classic ordered probit models for datasets of different sample sizes (*9*). The Bayesian ordered probit model produced better parameter estimates and prediction performance for small sample sizes with proper prior distribution setting. Various researchers have used mixed logit models (*10*) and ordered probit models (*11*) to conduct severity analysis. Another study by Chiou and Fu, which analyzed multi-period crash frequency and severity data, made use of an extended multinomial generalized Poisson model and spatiotemporal dependence (*12*).

Studies Based on Proportions of Severe Crashes

The MVPLN approach adopted by Ma et al. simultaneously modeled crash counts by injury severity (*8*). Parameter estimation within the Bayesian paradigm made use of a Gibbs sampler and Metropolis-Hastings algorithms to model crashes on rural highways for Washington State.

Park and Lord explored the flexibility of the MVPLN approach to cope with overdispersion and a fully general correlation structure in the data (*7, 8*) to estimate correlations across severity levels of multivariate crash counts from intersections in California. Chiou and Fu studied spatial and temporal dependence in both crash frequency and severity. The researchers used 5-year crash data at freeway segments in Taiwan to analyze extended multinomial generalized Poisson models with the spatial and spatiotemporal dependence functions. Although they found that temporal dependence is strong on crash frequency and is associated with traffic variability, they also found that crash severity exhibits strong spatial dependence and is associated with geometric configuration (*12*).

Council et al. developed severity models to study the effects of air bags on severity indexes for roadside objects and estimated the proportion of severe injuries experienced for various types of roadside hardware in fixed-object crashes (for type A + K crashes) (*13*). Classification and regression trees (CART method) were generated to control factors such as rural or urban crash location, speed limit, and vehicle type. Using the modeled proportions, the researchers concluded that an air bag reduced the severity of crashes by 10% to 50% (results based on $p \leq 0.10$), depending on the object. The beneficial effect was slightly greater for longitudinal barrier (guardrail) impacts compared with point object impacts. A larger sample size was recommended for better verification of air bag-related crash severity studies.

In 2015, Shea et al. studied changes in expected crash frequency and severity due to geometric design decisions on entrance–exit ramp spacing and auxiliary lane presence for 404 freeway segments in California and Washington (*2*). They developed SPFs using negative binomial (NB) regression models and SDF using MNL regression model predicted crash frequencies and estimated crash severities respectively. STATA software assisted in the estimation of severity models and the prediction of probability of crashes with a specific injury severity outcome. Expected crash frequencies increased for fatal (K), incapacitating injury (A), nonincapacitating injury (B), and possible injury (C) with decreasing ramp spacing. Auxiliary lanes were found to associate with decreased crash frequency but primarily associated with possible injury and property damage crashes (*2*).

Deng et al. analyzed the factors affecting the severity of head-on crashes on two-lane rural highways in Connecticut. Crashes involving two vehicles originally traveling in opposite directions were defined as head-on crash, thus opposite-direction sideswipe collisions and crashes involving turning vehicles were excluded. Ordered probit models were estimated to find the statistical association of head-on crash severity and various factors. The researchers found road surface condition (wet), pavement width (narrow), time of occurrence (night), high density of access points, and braking performance of the vehicles among such influential factors (*14*).

Crash and severity prediction models for Wisconsin freeways and interchanges, developed by Yu et al. to evaluate the effect of roadway lighting provided a base for benefit–cost analysis to prioritize freeways and interchanges for lighting installation systematically. The authors reduced the number of categorical and quantitative variables and used binary variables for all categorical factors to develop the SPFs. Six SPFs were generated for various lane-lighting combinations using a generalized linear model (GLM) with an NB distribution of crash frequency. The multinomial logistic model was selected for SDF development. Segment type and median width were significant factors in determining the crash severity of lighted segments, whereas for unlighted segments, segment type and international roughness index were significant factors (*15*).

Geedipally et al. developed SDFs for freeway segments to predict the proportion of crashes in each severity category as a function of roadway geometric design elements and traffic control features using California, Maine, and Washington data (*16*). Geedipally et al. used an MNL model to predict the probability of crash severities as well as the effects of barrier presence, high annual average daily traffic (AADT), rumble strips, horizontal curve, lane width, and area type. Installing the barrier reduced the likelihood of high severity crashes whereas presence of rumble strips increased them. Crashes were likely to be more severe on curves and less severe on roads with wider lanes. Urban areas reported less crash severity compared with rural areas. Another study (*3*) by the same authors analyzed the injury severity of crashes on ramp segments and signalized and unsignalized crossroad ramp terminals by developing separate SDFs using MNL models, which could be used in combination with SPFs and crash modification factors (CMFs) to obtain estimates of the safety effects of design decisions. State-specific variables were found to influence SDFs emphasizing the need for location calibration (*3*).

A probability-based severity model developed by Dixon et al. studied the influence of speed limit on total crashes and KAB crashes (KAB is an injury scale in which K = fatal [killed], A = incapacitating injury, B = nonincapacitating injury), at signalized intersections in Oregon (*17*). This SDF function was used to predict the proportion of KAB crashes that could be used in combination with the SPF for total crashes to produce KAB frequency predictions. The SPF for total crashes included both major and minor AADTs and SPF for KAB crashes, whereas the SDF included only terms for the speed limits on major and minor roads. The regression-based KAB frequency model also included terms for the two speed limits at the intersection.

Studies Based on Frequency Models for Crashes of Different Severities

More recently, Kitali et al. developed CMFs to quantify the impacts of pedestrian countdown signals (PCSs) to drivers using a before–after study with Empirical Bayes (EB) method. Full SPFs were estimated using NB regression for different crash types and severities. Crash frequencies were higher for commercial-type areas and intersections with higher speed limits (above 30 mph). The researchers found that PCSs associate with reductions of all type of crashes by between 4.6% and 8.8% (*18*).

In the framework of Poisson-lognormal mixture, a study by Dixon et al. developed SPFs for both total and KAB crashes (*19*). As mentioned earlier, the SPF for total crashes relied on both major and minor AADTs. Consistent with the SDFs (the probability-based model described in the previous subsection) the SPF for KAB crashes included both AADTs and terms for speed limits on the major and minor roads. However, the KAB frequency model was found to have a tendency to over-predict at higher-crash locations. It was argued that the small number of KAB crashes limits the ability to develop more robust frequency models. Finally, the combination of the SDF and the total crash SPF was recommended, which was developed based on a combined dataset (*17*).

Peel et al. analyzed the safety effectiveness of Wyoming snow fence implementations along a rural mountainous 19-mi section of Interstate 80 by using before–after analysis with EB utilizing Wyoming-specific SPFs as well as odds ratios (OR). This study further developed CMFs to estimate various crash types and severity levels. The authors also developed an OR for total crashes and fatal plus injury (F + I) crashes and estimated the OR in an effort to further understand and analyze the frequency of the crashes. The analysis encompassed 953 total winter crashes that occurred from 2003 to 2011. The EB method showed that all types of crashes tended to decrease with the presence of snow fences during the winter months (95% confidence level), regardless of weather conditions. F + I, property damage only crashes, and total crashes decreased by 59.1%, 23.2%, and 25.3%, respectively in all weather conditions (*20*).

Jestico et al. made use of a crowdsourced cycling incident dataset (2005–2016) from the Greater Victoria region, Canada to characterize the attributes of reported incidents at intersections between multiuse trails and roads; 98.1% of mapped incidents occurred between 2014 and 2016. The researchers used NB regression to model the relationship between the number of incidents and the infrastructure features. A higher proportion of incidents reported at multiuse trail-road intersections were collisions (versus near misses), with a higher proportion of incidents that resulted in injuries (35% versus 21%) (*21*). The major contributing factors were cycling volumes, vehicle volumes, and trail sight distance.

Similar to other efforts found in this literature review, Milton et al. proposed a combination of frequency models and the mixed logit model to determine severity proportions on overall roadway safety (*10*). This effort used crash data from the state of Washington at multilane divided highways.

Summary of Literature

The literature review found multiple applications that demonstrate the usefulness of both approaches of interest for this research. When large databases are available, the preference seems to be for disaggregating the data by crash severities and modeling the frequency of each subset of crashes separately (*18*, *20*, *21*). However, some researchers seem to prefer the development and use of SDFs even on larger datasets (*2*, *3*, *16*). The use of probability models, either as SDFs or as estimates of risk, appears to offer flexibility and a wide range of applications in safety evaluations (*3*, *13*).

Some works have contrasted the performance of predicting severe crash frequencies directly (via an SPF) or indirectly (via a combination of an SPF for total crashes and a SDF). Overall, research indicates a comparable performance in prediction by both approaches (*17*) and consistency in predictors from the KAB SPF approach and the total crashes SPF combined with SDF approach. The next section summarizes the databases available for a performance comparison of these competing approaches.

Alternative Methods of Analysis

This section summarizes the mathematical statistics related with the two types of analysis under evaluation in this paper. Although there are many alternatives within each of these types of analysis, the details provided in the next section correspond to models that were fitted to the two datasets that will be compared later in this paper.

Negative Binomial Regression for Crash Frequency

Under current practice, the statistical specification for crash frequency models assumes that the response variable (number of crashes) follows an NB distribution (*1*). Using this distribution to model the dispersion in the crash data allows more flexibility than assuming a more traditional Poisson distribution. The simplest of the NB specifications is that which treats the Poisson- overdispersion as a quadratic form of the expectation. This is simply known as the NB2 specification. For the NB2 specification, the modeling is performed again under the framework of GLMs. The link function is the natural logarithm in this case. The response variable is the observed number of crashes at every site in the analysis. The model then estimates the expected number of crashes as shown in Equation 1:

$$E(N|Site) = e^{X'\cdot\beta} \quad (1)$$

where

N = number of yearly crashes
$E(N|Site)$ = long-term expectation of N at a given site
$\boldsymbol{X}$ = vector of explanatory variables
$\boldsymbol{\beta}$ = vector of coefficients to be estimated

In addition to the parameterization of the expectation of crashes, the NB2 model requires the estimation of a dispersion parameter, such that the distribution of N follows the NB distribution. This parameter affects the variance of the number of crashes, as shown in Equation 2:

$$V(N|Site) = E(N|Site) + k \cdot E(N|Site)^2 \quad (2)$$

where $V(N|Site)$ = variance of the number of yearly crashes at a given site; and k = dispersion parameter; other variables as previously defined.

Variants from this basic definition for modeling data with uncommon characteristics exist. For example, recent research has proposed that the dispersion parameter of the NB distribution be parameterized, as dispersion in the data may vary with variables of interest, though there is concern that the need to parameterize the dispersion parameter may be a symptom of a poorly specified mean (*22*). For both datasets in this research, the authors found that either NB2 or Poisson crash distributions in a lognormal mixture sufficed to characterize the crash variability in each dataset.

Poisson Mixture Models

The NB distribution can be expressed as the mixture of a Poisson and a gamma distribution (*23*). In general, Poisson mixture models allow more flexibility to account for overdispersion with respect to the Poisson distribution. Estimation of most Poisson mixture models can also be carried out under the framework of GLMs. The link function is, as in the case of the NB2, the natural logarithm. The model then estimates the expected number of crashes, with the same specification previously shown in Equation 1. However, in contrast with the NB2 model, the variance of the number of crashes is not generally given in closed form. It can be shown (*24*) that the variance of a mixture should follow the relationship shown in Equation 3:

$$V(N) = E_{mixture}\left[V_{poisson}(N)\right] + V_{mixture}\left[E_{poisson}(N)\right] \quad (3)$$

Logistic Regression for Modeling Proportion of Severe Crashes

An alternative to crash frequency analyses is an analysis of proportions of a type of crash of interest. For this research, severe crashes are of interest, defined here as any crash of severity K, A, or B, in the KABCO scale (KAB plus C = possible injury, and O = property damage only). Under this approach, the response variable is the proportion of severe crashes, and the relationship with a potential explanatory variable is modeled as a change in the odds of a severe crash occurring. Similar to frequency analyses, this alternative requires having both severe and other types of crashes. Any co-dependency between data points is modeled as a set of random effects. A set of variables is defined, such that their effects are fixed and orthogonal to the random effects. The formal specification of the statistical model for KAB proportion of crashes is as follows:

$$Logit(\pi|Site) = \boldsymbol{Z}' \cdot \boldsymbol{\gamma} + \eta_j \quad (4)$$

where

$(\pi|Site)$ = proportion of KAB crashes at a given site

$Logistic(x)$ = logit function of x, defined as $\text{Logit}(x) = \ln\left(\frac{x}{1-x}\right)$

$\boldsymbol{Z}_{ik}$ = vector of k variables modeled as fixed effects over the dataset

η_j = random effect for a particular site j

$\boldsymbol{\gamma}$ = vector of coefficients for all variables in experiment (fixed effects, estimated)

In the above definition, it should be noted that the response variable is defined as the proportion of KAB crashes. The statistical specification links this proportion to a linear combination of the fixed effect variables.

When a proportional model is combined with a frequency estimate, the following assumption is implicit:

$$E(N_{KAB}) = E\left(\hat{N} \times \hat{\pi}_{KAB}\right) = E\left(\hat{N}\right) \times E(\hat{\pi}_{KAB})$$

where $\hat{N}$ and $\hat{\pi}_{KAB}$ are estimators of the parameters N and π_{KAB}.

However, this is only true if $\hat{N}$ and $\hat{\pi}_{KAB}$ are independent. Such independence is not assured if, for example, there are common covariates within the vectors of predictors $\boldsymbol{X}$ and $\boldsymbol{Z}$.

Analysis

This section summarizes two datasets and the corresponding models developed for estimation of KAB crashes. One dataset is from Oregon and one from Texas. The section also outlines the methodology for performing a sensitivity analysis on the performance of KAB predictions under various conditions. All analyses were performed with open source software and statistical packages (*25*, *26*).

Predictive Models from Oregon Intersection Sites

A set of three models was developed for safety performance of signalized intersections in Oregon (*17*). The SPFs and SDF from this project were developed using a Poisson-lognormal mixture model, both for total crashes and severe injury crashes (also coded as KAB). More details about this set of models can be found in the research technical report (*17*).

KAB Prediction Performance of Oregon Models. The coefficients from these models are shown in Table 1.

Some highlights of these models are as follows:

- The frequency model for total crashes included both major and minor AADTs
- The frequency model for KAB crashes included the same two covariates plus the speed limit on the major road
- The severity distribution model, predicting the proportion of KAB crashes, also included the speed limit, in addition to the speed differential between major and minor roads

To ease visual assessment of bias, Figure 1 shows two colors for points above or below the line of correspondence between the prediction and the actual number of crashes. In addition, this figure also shows the root mean squared error (RMSE) for each comparison. Figure 1 shows a relatively low range of KAB crashes. As a result, a graphical assessment of the performance in predicting KAB crashes is limited. It appears that both predictors have comparable performance in terms of RMSE and dispersion around the line of correspondence.

Table 1. Predictive Models for Total and KAB Crashes in Oregon (Signalized Sites)

Variables	Total crashes SPF	KAB crashes SPF	KAB crashes SDF
(Intercept)	−9.786 (1.717)***	−9.327 (2.978)**	−2.872 (0.515)***
Log (MjAADT)	0.708 (0.183)***	na	na
Log (MnAADT)	0.521 (0.131)***	na	na
Log (AADTMod)	na	0.754 (0.310)*	na
Major speed limit	na	0.041 (0.020)*	0.033 (0.015)*
Speed limit difference	na	na	−0.021 (0.011)†

Note: MjAADT = major AADT (veh/day); MnAADT = minor AADT (veh/day); AADTMod = module of AADTs (veh/day) defined as the squared root of the summed squared major and minor AADTs; Major speed limit = speed limit at major road (mph); Speed limit difference = difference in speed limits between major and minor roads (mph); na = not applicable.

†Statistically significant at the 10.0% significance level. *Statistically significant at the 5.0% significance level. **Statistically significant at the 1.0% significance level. ***Statistically significant at the 0.1% significance level.

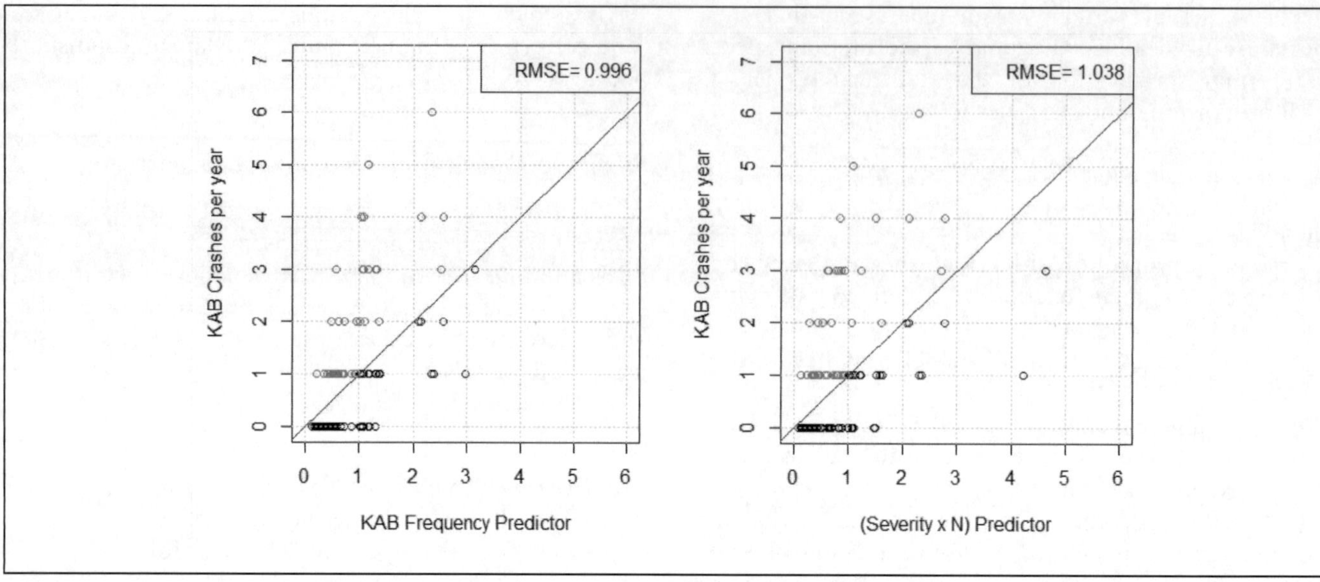

Figure 1. Model fits for alternative KAB predictors on Oregon data.

Table 2. Predictive Models for Total and KAB Crashes in Texas (Signalized Sites)

Variables	Total crashes SPF	KAB crashes SPF	KAB crashes SDF
(Intercept)[a]	5.304 (1.086)***	3.076 (1.110)†	−1.039 (0.048)***
Scale (cross street AADT)	0.113 (0.049)*	na	na
Ln (cross street AADT)	na	0.072 (0.039)†	na
AvgLn	0.703 (0.162)***	0.480 (0.158)**	na
Ln (Rmin)	−0.951 (0.245)***	−0.668 (0.241)**	na
Scale (DWY)	−0.268 (0.072)***	−0.197 (0.070)**	na
Right turn from cross road merges into frontage	na	−0.180 (0.089)*	na
Right turn shared lane from cross road	−0.271 (0.102)**	−0.306 (0.101)**	na
Scale (average spacing between frontages)	na	na	0.165 (0.047)***
Scale (average taper opening)	na	na	−0.107 (0.048)*
Unequal posted speed limits	na	na	0.739 (0.206)***

Note: na = not applicable.
†Statistically significant at the 10.0% significance level. *Statistically significant at the 5.0% significance level. **Statistically significant at the 1.0% significance level. ***Statistically significant at the 0.1% significance level.

Predictive Models from Texas Turnaround Sites

A set of three models was developed to assess the safety performance of U-turns at frontage roads turnarounds in Texas. These models are based on crash data for years 2009 to 2015 and site data collected using aerial imagery (*27*). The, dataset consisted of signalized sites with and without U-turns. The authors performed a stepwise regression analysis to assess the influence of significant variables on the total number of crashes as well as crash severity for the dedicated U-turn lane locations with a posted speed limit of at least 30 mph. The crash model development focused on the identification of statistically significant variables and an assessment of the best functional form that represents each identified variable. It was noted that, although three AADT values were available (one for each frontage road and one for the cross-road) only the cross road AADT was retained in the models. The authors speculate that the reliability of the frontage road AADT may explain why these values were ultimately excluded in the model selection.

KAB Prediction Performance of Texas Models. After an iterative process of model selection, considering various variables and subsets of data, the research team arrived to the most parsimonious frequency models for total crashes and KAB crashes. The research team also developed a severity distribution function to be utilized in combination with the SPF developed for total crashes as

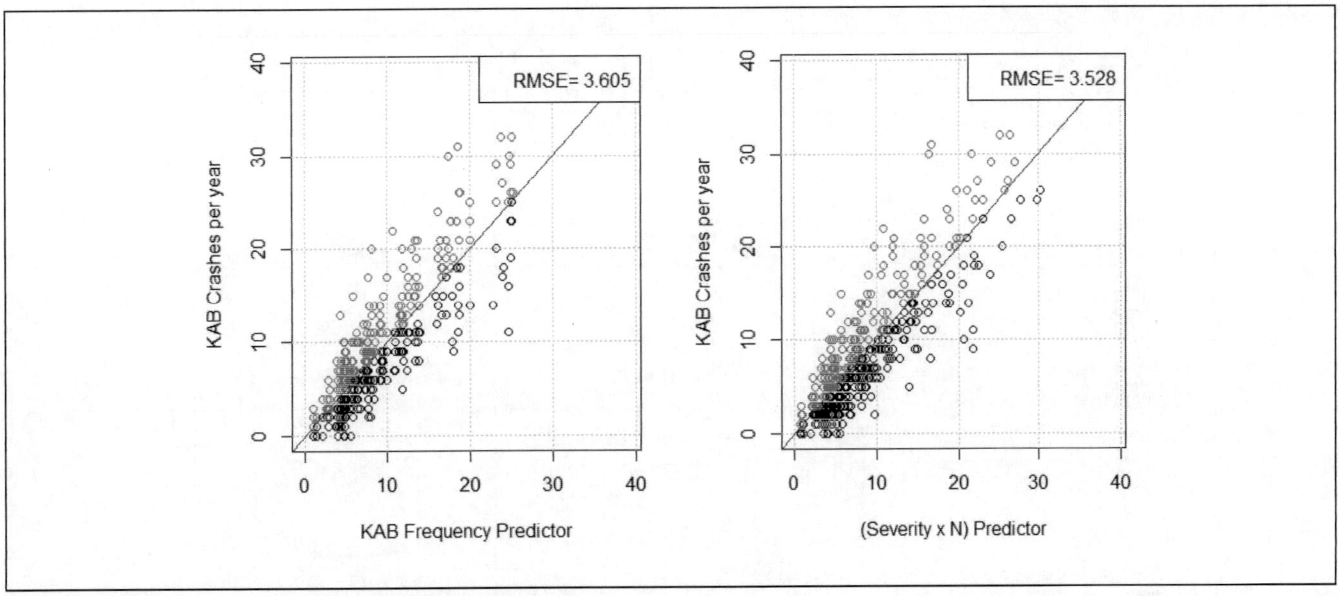

Figure 2. Model fits for alternative KAB predictors on Texas data.

an alternative to the frequency model for KAB crashes. The coefficients for these models are shown in Table 2.

Many of the predictors are expectedly in common between the total crashes SPF and KAB crashes SPF. In addition, the signs and magnitudes of effects agree between the two SPFs. However, it can be noted that the SDF has a completely different non-overlapping set of predictors, compared with the two SPFs.

Next, the research team performed a comparison between the two ways of estimating KAB crashes: (1) the frequency predictor of KAB, developed from the SPF for KAB crashes; and (2) the proportional predictor of KAB, which combines the result from the SPF for total crashes and the KAB SDF. Results from this comparison are shown in Figure 2.

An examination of Figure 2 indicates that, similar to Oregon, the two predictors in Texas are free of bias, with similar dispersions and with very comparable RMSE values.

Summary of Comparisons of KAB Prediction Performance in Oregon and Texas

Both Figure 1 and Figure 2 indicate very similar performance in KAB prediction. This similarity is insightful, given the dissimilarities between the models, particularly:

- The Oregon models are for street intersections, whereas the Texas models are for frontage road turnaround locations, whose design and operation are more complex. Consequently, the range of KAB crashes is narrow in Oregon but wider in Texas.
- The dispersions for the frequency models are functionally different for both sets.
- The amount of overlap of predictors between the KAB SPF and the SDF is significant in the Oregon models, but inexistent in the Texas models.
- When considering the combination of total crashes SPF and SDF jointly and contrasting it with the KAB SPF, both the Oregon and Texas models tended to show a large set of common variables among their predictors.

The following section of the paper describes a sensitivity analysis on synthetic data in order to explore boundary conditions for the performance of both types of estimation.

Sensitivity Analysis

For this sensitivity analysis, batches of 1,000 replications of synthetic data were generated from a NB distribution to approximate the distribution of KAB crashes (and corresponding total crashes) observed in the Texas dataset. The three types of models discussed in this paper were fitted to each synthetic sample and the corresponding pair predictors were obtained. In order to observe the effects of varying certain parameters, a set of progressive and cumulative variations were introduced to each additional batch of synthetic data.

Evaluations in Synthetic Data. The research team prepared a set of tests of KAB prediction performance to be

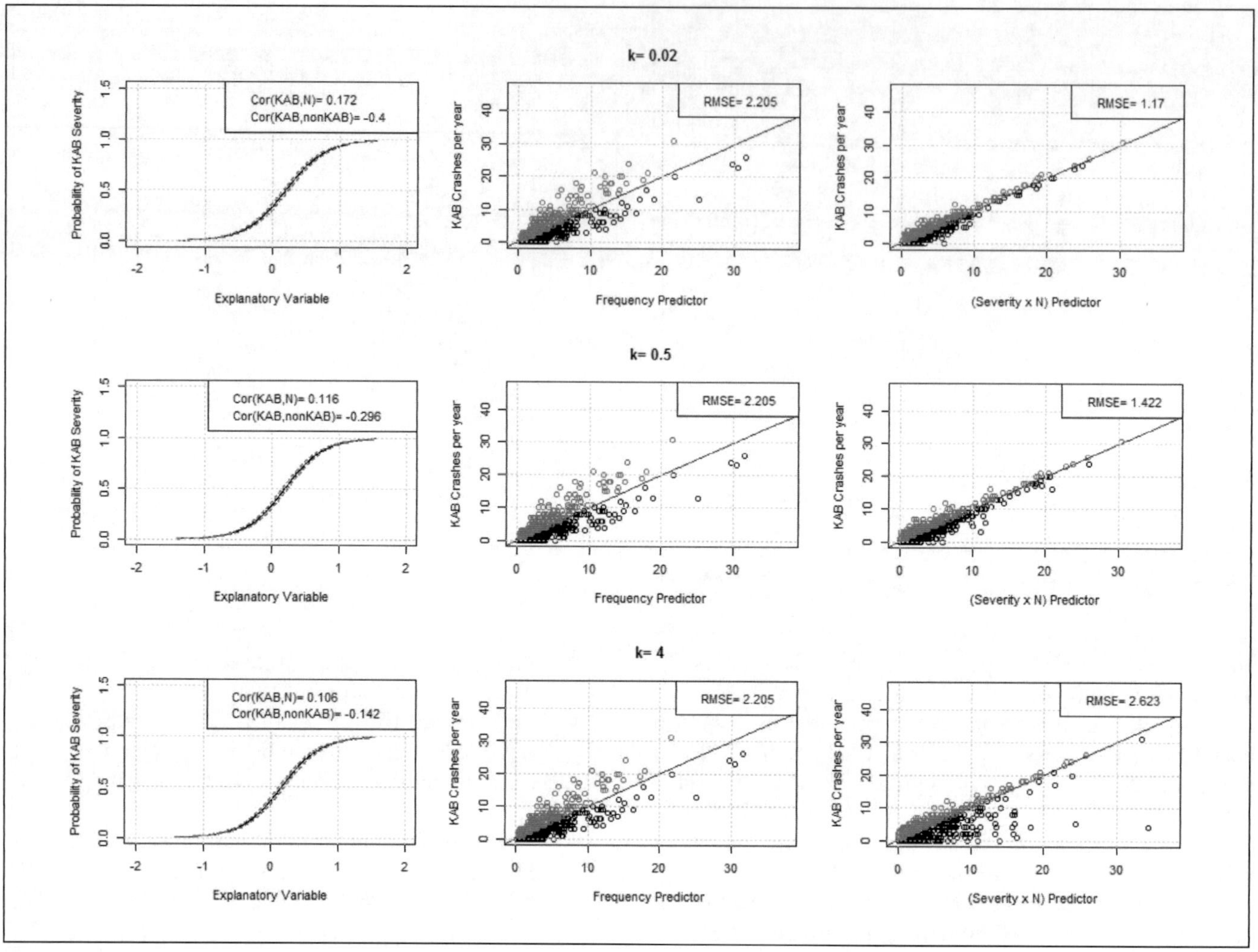

Figure 3. Model performance by increasing dispersion.

evaluated using synthetic data. The intent was to characterize how KAB prediction varies with increasing dispersion in the data, with different rates of growth for KAB and non-KAB crashes, and in relation to decreasing kurtosis in the SDF curve. In contrast with the performance evaluations in the previous sections, this sensitivity analysis adopted a validation scheme: two synthetic datasets were used, one to fit the frequency and severity models and another to test their performance in prediction. Figure 3 shows a battery of tests with different dispersion parameters. It can be seen in this figure that the performance of the combination of SDF and total crashes SPF deteriorates with increasing dispersion.

Similar to Figure 3, Figure 4 shows the performance of predictions at different rates of increase for KAB crashes, compared with non-KAB crashes. This figure indicates that the performance of the combination of SDF and total crashes SPF is better than the KAB frequency predictor and relatively invariant when KAB crashes increase at the same or a higher rate as non-KAB crashes. However, the performance of the indirect predictor (SDF $\times$ SPF_{total}) starts to diminish if non-KAB crashes happen to increase at a higher rate than KAB crashes.

Finally, Figure 5 shows the performance of the predictors at different levels of correlation between KAB and non-KAB crashes. Again, the performance of the indirect predictor (SDF $\times$ SPF_{total}) is better than the KAB frequency direct predictor when the correlation between KAB and non-KAB crashes is either positive or null and it deteriorates rapidly when this correlation is negative.

Conclusions

This research investigated the tradeoffs of two approaches to model severe crashes: a direct approach,

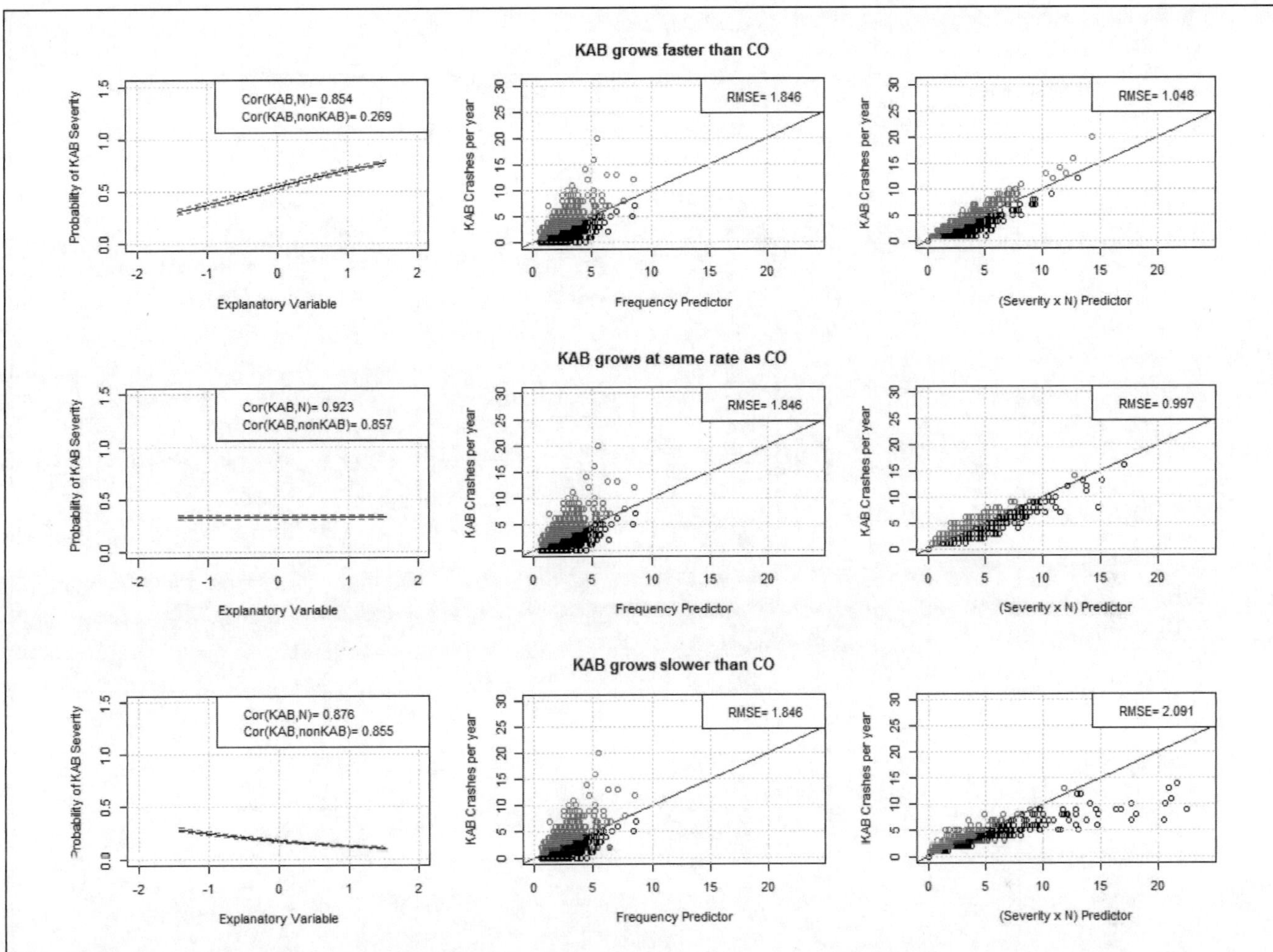

Figure 4. Model performance by relative rate of increase of KAB crashes.

which consists in developing a model for the severe crashes; or an indirect approach, which consists in combining the result of an SPF for total crashes with the estimation of the proportion of severe crashes.

Two datasets were used to compare the performance of these approaches and it was found that the performance of each approach is on a par with each other. It was noted that such comparable performances from such different states and samples of sites might indicate robustness of the methods to some study characteristics.

The research team performed a sensitivity analysis on synthetic data and it was found that, in terms of dispersion around the prediction, the indirect approach (i.e., $SDF \times SPF_{total}$) outperforms the direct approach (i.e., prediction from severe crashes SPFs) most of the time. However, these comparisons found that the indirect approach had poor performance under the following conditions: increased dispersion, a lower rate of increase of KAB crashes (compared with non-KAB crashes), and when KAB crashes and non-KAB crashes tend to correlate negatively.

Future Work

The following list identifies future work the authors believe is still needed under the topic of the performance of alternative methods to model crash severity:

- Research is needed to help researchers or practitioners identify what types of real crash data displays the conditions this work identified that would result in issues for one method or the other.
- Implementation work is needed that develops guidelines to help practitioners choose one approach over the other.
- Finally, more research is needed to test a wider range of conditions for comparison of prediction performance.

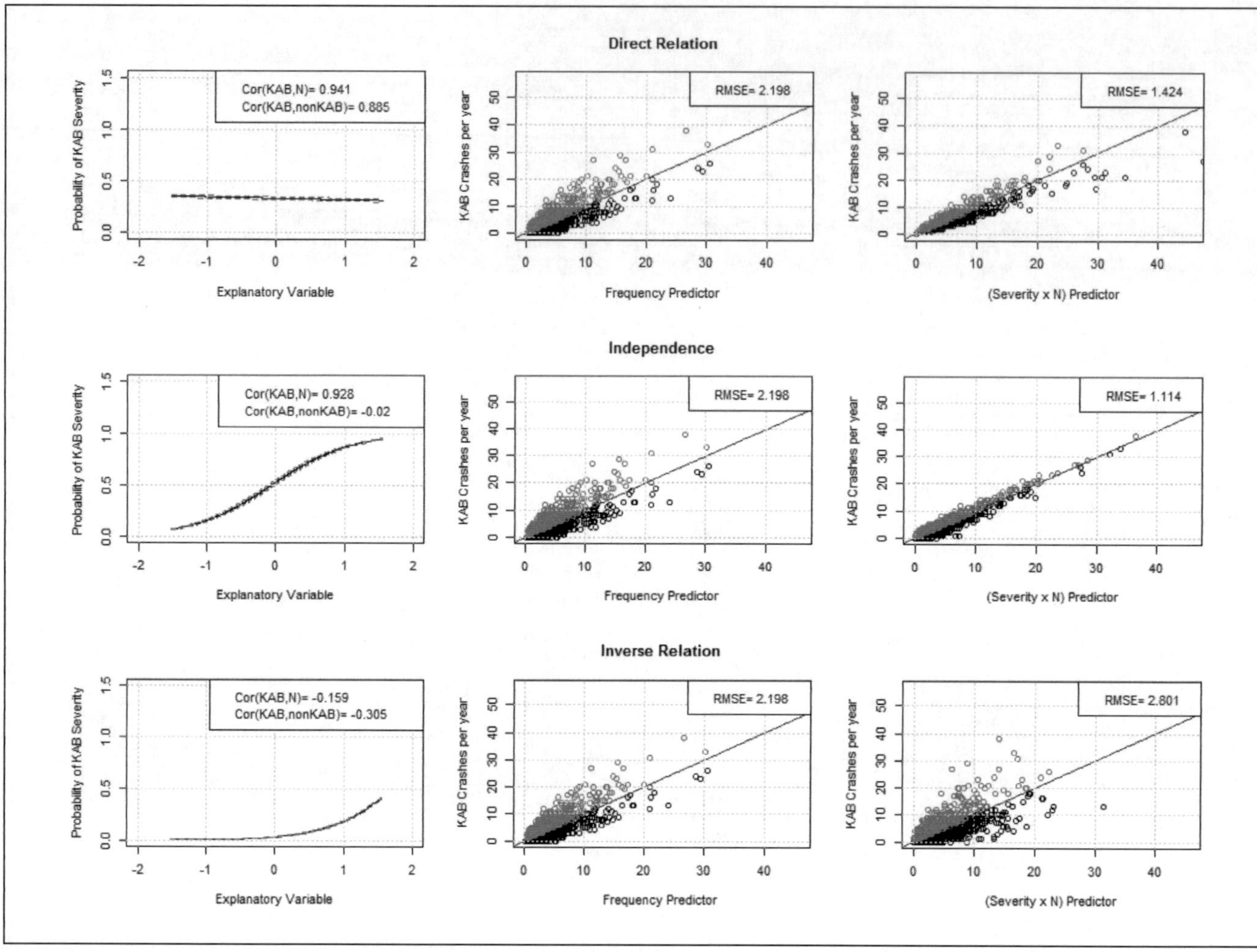

Figure 5. Model performance by correlation between KAB and non-KAB crashes.

Acknowledgments

This research project was supported by the Texas A&M Transportation Institute Center for Transportation Safety. The research team also acknowledges the Oregon Department of Transportation and the Texas Department of Transportation projects, as data from projects funded by these agencies were used in this study. The authors also thank TTI staff members for their assistance in collecting, reducing, and filtering field data, and offering useful suggestions on the manuscript.

Author Contributions

The authors confirm contribution to the paper as follows: study conception and design: RA, KD; data collection: RA, SA; analysis and interpretation of results: RA, KD; draft manuscript preparation: RA, SA. All authors reviewed the results and approved the final version of the manuscript.

References

1. *Highway Safety Manual.* AASHTO, Washington, D.C., 2010.
2. Shea, S., T. Le, and R. Porter. Combined Crash Frequency–Crash Severity Evaluation of Geometric Design Decisions: Entrance–Exit Ramp Spacing and Auxiliary Lane Presence. *Transportation Research Record: Journal of the Transportation Research Board*, 2015. 2521: 54–63.
3. Geedipally, S., J. Bonneson, M. Pratt, and D. Lord. Analysis of Injury Severity in Crashes on Ramps and at Crossroad Ramp Terminals. *Transportation Research Record: Journal of the Transportation Research Board*, 2014. 2435: 37–44.
4. Lord, D., S. Geedipally, and M. Shirazi. *Improved Guidelines for Estimating the Highway Safety Manual Calibration Factors.* Publication ATLAS-2015-10. Advancing Transportation Leadership and Safety (ATLAS) Center, Ann Arbor, MI, 2015.
5. Savolainen, P., F. Mannering, D. Lord, and M. Quddus. The Statistical Analysis of Highway Crash Injury Severities: A Review and Assessment of Methodological Alternatives. *Accident Analysis and Prevention*, Vol. 43, No. 5, 2011, pp. 1666–1676.
6. Lord, D., and F. Ye. Comparing Three Commonly Used Crash Severity Models on Sample Size Requirements:

Multinomial Logit, Ordered Probit and Mixed Logit Models. *Analytic Methods in Accident Research*, Vol. 1, 2014, pp. 72–85.

7. Ma, J., K. Kockelman, and P. Damien. A Multivariate Poisson-Lognormal Regression Model for Prediction of Crash Counts by Severity using Bayesian Methods. *Accident Analysis and Prevention*, Vol. 40, 2008, pp. 964–975.
8. Park, E., and D. Lord. Multivariate Poisson–Lognormal Models for Jointly Modeling Crash Frequency by Severity. *Transportation Research Record: Journal of the Transportation Research Board*, 2007. 2019: 1–6.
9. Xie, Y., Y. Zhang, and F. Liang. Crash Injury Severity Analysis Using Bayesian Ordered Probit Models. *Journal of Transportation Engineering*, Vol. 135, No. 1, 2009, pp. 18–25.
10. Milton, J., V. Shankar, and F. Mannering. Highway Accident Severities and the Mixed logit Model: An Exploratory Empirical Analysis. *Accident Analysis and Prevention*, Vol. 40, 2008, pp. 260–266.
11. Kockelman, K., and Y. Kweon. Driver Injury Severity: An Application of Ordered Probit Models. *Accident Analysis and Prevention*, Vol. 34, 2002, pp. 313–321.
12. Chiou, Y., and C. Fu. Modeling Crash Frequency and Severity with Spatiotemporal Dependence. *Analytic Methods in Accident Research*, Vol. 5–6, 2015, pp. 43–58.
13. Council, F. M., Y. Mohamedsha, and J. Stewart. Effects of Air Bags on Severity Indexes for Roadside Objects. *Transportation Research Record: Journal of the Transportation Research Board*, 1997. 1581: 66–71.
14. Deng, Z., J. Ivan, and P. Garder. Analysis of Factors Affecting the Severity of Head-On Crashes: Two-Lane Rural Highways in Connecticut. *Transportation Research Record: Journal of the Transportation Research Board*, 2006. 1953: 137–146.
15. Yu, L., Z. Li, A. Bill, and D. Noyce. Development of Freeway and Interchange Safety Performance Functions with Respect to Roadway Lighting: A Pilot Study. *Transportation Research Record: Journal of the Transportation Research Board*, 2015. 2485: 16–25.
16. Geedipally, S., J. Bonneson, M. Pratt, and D. Lord. Severity Distribution Functions for Freeway Segments. *Transportation Research Record: Journal of the Transportation Research Board*, 2013. 2398: 19–27.
17. Dixon, K., C. Monsere, R. Avelar, J. Barnett, P. Escobar, S. Kothuri, and Y. Wang, *Improved Safety Performance Functions for Signalized Intersections*. Oregon Department of Transportation, Salem, OR, 2015.
18. Kitali, A., T. Sando, A. Castro, D. Kobelo, and J. Mwakalonge. Developing Crash Modification Factors to Quantify Impacts of Pedestrian Countdown Signals to Drivers. Presented at 96th Annual Meeting of the Transportation Research Board, Washington D.C., 2017.
19. Dixon, K., R. Avelar, and C. Monsere. Oregon Signalized Intersection Safety Performance Functions and the Effect of Speed. Presented at 95th Annual Meeting of the Transportation Research Board, Washington, D.C., 2016.
20. Peel, T., and M. Ahmed. Investigating Safety Effectiveness of Wyoming Snow Fence Installations Along a Rural Mountainous Freeway. *Transportation Research Record: Journal of the Transportation Research Board*, 2017. 2613: 8–15.
21. Jestico, B., T. Nelson, J. Potter, and M. Winters. Multiuse Trail Intersection Safety Analysis: A Crowdsourced Data Perspective. *Accident Analysis & Prevention*, Vol. 103, 2017, pp. 65–71.
22. Mitra, S., and S. Washington. On the Nature of Over-Dispersion in Motor Vehicle Crash Prediction Models. *Accident Analysis and Prevention*, Vol. 39, No. 3, 2007, pp. 459–468.
23. Hilbe, J. M. *Negative Binomial Regression*, 2nd ed. Cambridge University Press, New York, NY, 2011.
24. Wackerly, D., D. Mendenhall, III, and R. Scheaffer. *Mathematical Statistics with Applications*, 7th ed. Thomson, Toronto, Canada, 2008.
25. R Core Team. *R: A Language and Environment for Statistical Computing*. R Foundation for Statistical Computing, Vienna, Austria, 2015.
26. Bates, D., M. Mächler, B. Bolker, and W. S. Fitting. Linear Mixed-Effects Models using lme4. *Journal of Statistical Software*, Vol. 67, No. 1, 2015, pp. 1–48.
27. Google Inc. *Google Earth Pro v7.1.7.2606*. Mountain View, CA, 2016.

The Standing Committee on Highway Safety Performance (ANB25) peer-reviewed this paper (18-00495).

Article

TRR
JOURNAL OF THE TRANSPORTATION RESEARCH BOARD

Transportation Research Record
2018, Vol. 2672(30) 120–130
© National Academy of Sciences:
Transportation Research Board 2018
Article reuse guidelines:
sagepub.com/journals-permissions
DOI: 10.1177/0361198118796000
journals.sagepub.com/home/trr
SAGE

Safety Analysis of Interchange Functional Areas

Akinfolarin Abatan[1] and Peter T. Savolainen[2]

Abstract

Limited access facilities, such as freeways and expressways, are generally designed to the highest standards among public roads. Consequently, these facilities demonstrate crash, injury, and fatality rates that are significantly lower than other road facility types. However, these rates are generally elevated in the immediate vicinity of interchanges because of increases in traffic conflicts precipitated by weaving, merging, and diverging traffic. Given the extensive costs involved in interchange construction, it is important to discern the expected operational and safety impacts of various design alternatives. To this end, the objective of this study was to analyze safety performance within the functional areas of interchanges. The study involves the integration of traffic crash, volume, and roadway geometric data from 2010 to 2014 in the state of Iowa. Separate analyses were conducted for the freeway mainline and ramp connections. A series of safety performance functions (SPFs) were estimated for both the mainline and ramps. Random effects negative binomial models were estimated, which account for correlation in crash counts at the same location over time. The results show the frequency of crashes to vary based on traffic volume, interchange configuration, speed limit, and traffic control at the ramp terminal. The random effects models are shown to significantly outperform pooled models, which suggest there are several important location-specific factors that are not included in the analysis dataset. The SPFs from this study are also compared with several reference models from the extant research literature.

Interchanges are critical to long-term access management policies. Effectively designed interchanges allow for traffic and land use control that optimize both traffic safety and operations. There are an estimated 17,800 interchanges on the U.S. Interstate Highway System and about 6,900 on other access-controlled highways. Less than 35% of these interchanges are also located in more rural areas (*1*). As the U.S. freeway system ages and becomes more congested, many existing systems will need rehabilitation, redesign, improvement, or even reconstruction (*2*). Many interchanges provide freeway access to other high-volume roads with more limited access control. Therefore, interchanges present potential points of conflict on such roads. There is evidence that crash risk is significantly heightened on freeways in the interchange functional area as compared with areas upstream or downstream of this area (*3*). Opportunities for conflicts and, consequently, crashes are usually also more prominent in closer proximity to interchanges. Consequently, the development of crash prediction models specific to interchanges is expected to provide better predictive capabilities as compared with broader, aggregate-level models that analyze freeways without regard to the nuances of functional area. A better or more reasonable approach may therefore involve analyzing interchange functional areas as a separate explicit facility type (*4*).

To this end, the Interchange Safety Analysis Tool (ISAT) was designed by the Federal Highway Administration (FHWA) to carry out safety assessments of freeway-to-freeway and freeway-to-arterial interchanges. ISAT includes analysis modules for mainline freeway segments, interchange ramps, and crossroad (non-freeway) segments. ISAT makes use of safety performance functions (SPFs) developed for the SafetyAnalyst software and based on data from California, Minnesota, Ohio, and Washington. Therefore, to provide directly applicable results for other states, calibration coefficients based on state-specific data are recommended (*5*). Similar calibration is recommended for the SPFs for freeways and interchanges that were developed as a part of NCHRP 17-45, which were

[1]Fehr and Peers, Walnut Creek, CA
[2]Department of Civil and Environmental Engineering, Michigan State University, East Lansing, MI

Corresponding Author:
Address correspondence to Peter T. Savolainen: pete@msu.edu

included as a supplement to the first edition of the Highway Safety Manual (HSM) (*6*) and integrated into the Interchange Safety Analysis Tool-Enhanced (ISATe), which was released in 2012. One practical limitation to the use of the SPFs from either ISAT or the HSM/ISATe is the amount of data required for calibration. For example, the predictive method for basic freeway segments in the HSM requires a total of 14 data elements whereas the method for ramps requires 10 elements. Unfortunately, several of these elements require a high level of effort as noted in NCHRP 17-45 (e.g., length and radii of horizontal curves, length of/offset to median barriers, clear zone width), which can inhibit the effective utilization of these tools. In addition, a recent meta-analysis of SPFs for freeway merge and diverge areas finds that existing research in this area has been largely inconsistent (*7*). The authors note the challenges in transferability of models to other contexts, which may be reflective of differences in geometric characteristics in the areas where these studies have been conducted.

An attractive alternative to using calibrated SPFs from ISAT to carry out safety analyses for interchange areas is the development of jurisdiction-specific SPFs for these facility types. Jurisdiction-specific SPFs also provide the opportunity to examine additional characteristics (depending on the availability of data) that may not be possible with a calibrated SPF. The other alternative to analyze the functional area of interchanges would be operational knowledge of ISAT to calibrate the application for a particular jurisdiction.

In a study to examine the characteristics of crashes at interchange areas, it was observed that interchange-related fatal crashes constituted 22% of fatal crashes on or related to the freeway system. Thus, interchange-related crashes represented a substantial percentage of fatal crashes occurring on freeways. The results from this study estimate that each interchange experiences, on average, 0.05 fatal crashes per year and about 12.5 total crashes per year when considering all crash severity levels (*2*). As engineers develop new and unconventional designs, like the folded interchange, and aging infrastructure on state-maintained roads is upgraded to resolve present or potential issues, it will be beneficial to be able to quantify the relative safety of various interchange configurations. (*8*).

Literature Review

Safety on interchanges and within the interchange functional area has long been an area of concern for civil engineering designers and researchers because of elevated crash rates in the immediate vicinity of interchanges compared with the rest of the freeway, rates which are in turn caused by the higher amount of conflict introduced by merging, diverging, and weaving traffic. In a study to compare crash rates for interchange corridors with the average rates of adjacent corridors, it was observed that the interchange corridors experienced higher comparative crash rates. The interchange corridors were generally different than the adjacent corridors and provided additional complexities for drivers in that they presented additional conflict points, as well as potential confusion and more difficult decision making, which resulted in higher overall crash rates (*9*).

According to the ISAT user manual, the interchange influence areas (IIA) for mainline freeway segments are defined to extend 0.3 mi upstream from the gore point of the first ramp of a particular interchange to approximately 0.3 mi downstream from the last ramp of a given interchange. Conversely, all freeway segments beyond this limit are assumed not to be under the influence of the interchange conditions and safety effects of weaving are expected to have been completely dissipated (*5*).

Many states currently do not explicitly define the IIA in their roadway inventory database. The extent of influence area of the interchange to freeway within also seems to be logical rather than empirical. In a study conducted to develop state-specific SPFs for deployment of SafetyAnalyst across the state of Florida, 0.3 mi was considered as the IIA cut-off point. In this study, the IIA freeway segments were separated from other freeway segments by creating a 0.3-mi buffer for each ramp of the interchange (*10*). In another study in North Carolina to develop safety prediction models for influence areas of freeway ramps, crashes that occurred in speed-change lanes on the freeway up to 1,500 ft from the gore point were classified as freeway ramp-related crashes (*11*).

Interchange rehabilitation and reconstruction projects can especially benefit from the development of accurate interchange segment SPFs. As highlighted by the HSM, one of the major applications of SPFs is in determining the safety impacts of roadway design changes at a project level. Part C of the HSM provides methods for evaluation of proposed freeway and ramp design conditions against existing conditions. The ISAT is an example of an end-product FHWA tool based on data from four states in the U.S. that can be used to carry out crash analysis at the influence area of interchanges. However, SPFs from the tool are only valid for application to states and time periods for which the models were developed. The SPFs were developed using available data from the FHWA Highway Safety Information System (HSIS) for California (1997–2001), Minnesota (1995–1999), Ohio (1997–1999), and Washington (1993–1996) (*2*). ISAT has an input for calibration coefficients that allow calculations performed within ISAT to reflect better local safety experience. In a study that compared locally developed SPFs for rural two-lane highways in

Kansas with the default SPFs in the HSM that were calibrated for the state of Kansas, the resulting SPFs were found to be similar (*12*). However, in another study conducted in Florida that compared locally developed SPFs with ones developed by calculating calibration factors for the SafetyAnalyst SPFs, the Florida-specific SPFs were found to provide better fit than the SafetyAnalyst default SPFs calibrated to Florida data in most cases (*13*). The results from this study therefore suggest that agencies could benefit from developing their own local or state-specific SPFs.

Research Objectives

Given that highway interchanges are a complex part of the interstate system, which have received only limited coverage in statewide safety analyses, the objective of this study is to carry out a series of statistical analyses focused on interchange functional areas to determine the relationship between various operational and geometric roadway factors on crashes within the functional area. Various roadway databases, a crash database, and other resources maintained by the Iowa Department of Transportation (DOT) were leveraged as a part of this study. Initially, a list of about 340 interchanges in Iowa provided a starting point, which was subsequently expanded to include a total of 423 interchanges in this statewide crash analysis.

Empirical crash analyses of the IIAs for various interchange configurations were conducted to contrast their safety performance. A series of SPFs, in the form of generalized linear regression models, were estimated to determine those factors associated with crashes on interchanges based on a physical classification of interchange segments. Different interchange configurations were investigated to determine whether any specific configurations are associated with better, or worse, safety performance.

Data Description

For the purpose of this study, a series of Iowa DOT maintained databases were integrated to carry out an exploratory crash analysis and develop SPFs. The Iowa DOT GIMS database includes a series of GIS shapefiles that contain statewide information detailing roadway geometric features and traffic operational characteristics, which are georeferenced and hosted on an Iowa DOT public server. In this study, the GIMS data were extracted for the 5-year analysis period from 2010 to 2014. The project also utilized the Iowa DOT intersection database, which contains information about all roadway intersections on state-maintained roads in Iowa. This comprehensive database also contains information on ramp terminal control. Information about ramp advisory speeds was obtained from the Iowa DOT sign inventory. The Iowa DOT also maintains a crash database that is updated annually to include all crashes reported on Iowa roads. This data is based on crash report forms collected by law enforcement investigating officers. The crash-level file contains information about the universal characteristics associated with each crash, such as the crash type, location, and severity level. For this study, 423 interchanges in the state of Iowa were considered. The tasks involved as a part of the data collection process are briefly summarized here:

- Identification of the physical location of all interchanges in the state of Iowa.
- Collection of GIMS data for each interchange in the study area in GIS through a series of GIS filters and manual review.
- Classification of each segment based on facility type and interchange configuration.
- Truncation of crossroad GIMS segments at the IIA ramp terminals and freeway mainline GIMS segments at distances approximately 0.25 mi from each ramp terminal (see Figure 1).
- Association of crashes occurring from 2010 to 2014 within the vicinity of each interchange with the appropriate GIMS segment in the IIA, using a spatial selection tool in GIS to attach crashes to the mainline, ramp, or crossroad, depending on which of these features the crash was located nearest to.

After the completion of the joining process, the data was aggregated into a single database for the mainline and another database for the ramps that contained relevant site characteristics of interest for analysis. Table 1 provides a summary of the data for the interchange mainline. About 40% of the interchange mainlines are in urban area boundaries as designated by the FHWA. Half of the interchange mainlines in Iowa are operating under level of service (LOS) A whereas nearly 49% of the interchange mainlines are operating under LOS B. Table 2 provides a summary of the data for the interchange ramps, with this dataset including information for 1,625 ramp analysis segments that were investigated.

Statistical Methodology

Statistical investigation was carried out to determine the variables associated with the frequency of crashes occurring in the vicinity of interchanges. Areas of concern examined as a part of this investigation included the freeway mainlines, the cross-streets, and the ramp areas, with particular interest on the signal control at the ramp

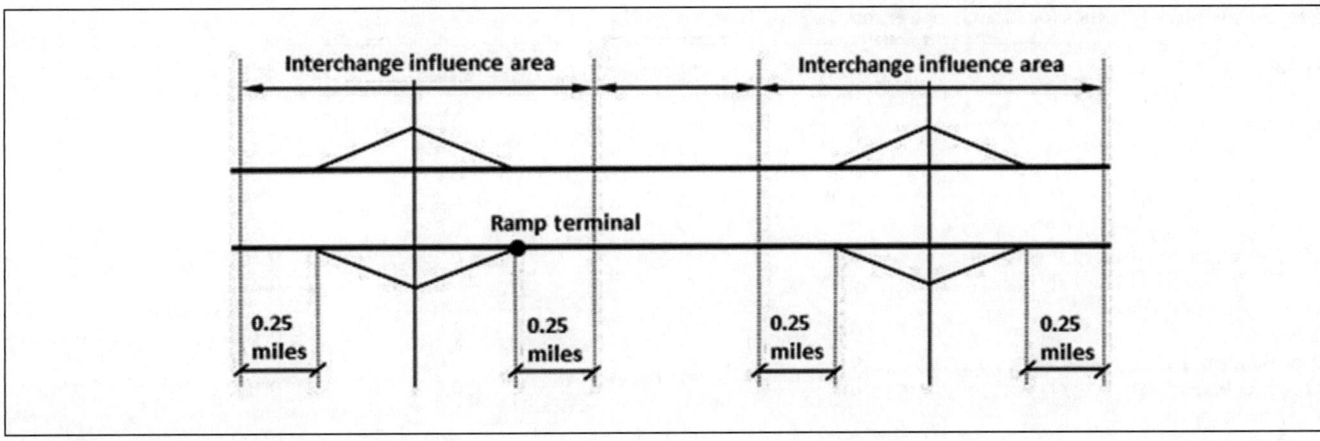

Figure 1. Interchange influence area (IIA).

Table 1. Summary Statistics for Interchange Mainline Data

Parameter	Minimum	Maximum	Mean	Std. dev.	No. of segments
Total crashes	0	54.00	6.01	6.96	2,015
Annual average daily traffic (AADT)	1,735.71	121,812.50	21,609.39	19,977.67	2,015
Mainline length (mi)	0.23	5.07	1.29	0.55	2,015
Speed limit					
Less than 55 mph	0	1	0.02	0.13	25
55 mph	0	1	0.20	0.40	410
60 or 65 mph	0	1	0.47	0.5	940
70 mph	0	1	0.32	0.47	640
Interchange configuration					
Diamond	0	1	0.69	0.46	1,395
Partial cloverleaf	0	1	0.24	0.43	480
Trumpet	0	1	0.03	0.16	50
Any directional	0	1	0.02	0.15	45
Full cloverleaf	0	1	0.02	0.15	40
Single-point urban interchange	0	1	<0.01	0.05	5
Locality					
Urban	0	1	0.40	0.49	810
Rural	0	1	0.60	0.49	1205
Number of lanes					
4 lanes	0	1	0.89	0.31	1,805
6 lanes	0	1	0.08	0.27	165
8 lanes	0	1	0.02	0.13	35
Other/unknown	0	1	0.01	0.09	10
Design level of service					
LOS A	0	1	0.50	0.50	1,005
LOS B	0	1	0.48	0.49	975
LOS C–D	0	1	0.01	0.1	20
Other/uknown	0	1	0.01	0.1	15

Note: Std. dev. = standard deviation; No. = number; AADT = annual average daily traffic.

terminals. Because interchange configurations have different geometric designs that may influence crashes, the extent of this influence was also investigated.

First, an exploratory analysis of the dataset was conducted to determine general trends, discern sample sizes within various categories of interest, and provide descriptive statistics of the datasets used. SPFs were developed to relate the frequency of crashes within the interchange functional areas to a series of observable roadway characteristics. For traffic safety data, SPFs generally include annual average daily traffic (AADT) and length in the case of segment SPFs. These models

Table 2. Summary Statistics for Interchange Ramp Data

Parameter	Minimum	Maximum	Mean	Std. dev.	No. of ramps
Total crashes	0	20	0.57	1.31	8,125
Annual average daily traffic (AADT)	5	27,450	1,806.49	2,316.97	8,125
Ramp length (mi)	0.04	1.09	0.28	0.09	8,125
Ramp advisory speed					
No ramp advisory speed	0	1	0.86	0.35	6,960
10–35 mph ramp advisory speed	0	1	0.08	0.26	610
40–55 mph ramp advisory speed	0	1	0.07	0.25	555
Number of lanes on ramp					
1 lane	0	1	0.96	0.20	7,835
2 or 3 lanes	0	1	0.03	0.17	215
Other/unknown	0	1	0.01	0.10	74
Ramp geometry					
Diagonal ramp	0	1	0.87	0.34	7,010
Loop ramp	0	1	0.13	0.34	1,030
Other	0	1	0.01	0.10	85
Type of ramp					
Freeway-to-freeway ramp indicator	0	1	0.05	0.10	410
On-ramp indicator	0	1	0.48	0.50	3,890
Off-ramp indicator	0	1	0.47	0.50	3,825
Presence of signal by ramp type					
Signalized on-ramp	0	1	0.06	0.24	480
Signalized off-ramp	0	1	0.08	0.28	680
Unsignalized ramp	0	1	0.86	0.35	6,965
Design level of service (LOS)					
LOS A	0	1	0.51	0.50	4,130
LOS B	0	1	0.42	0.49	3,385
LOS C–D	0	1	0.01	0.12	180
Other/unknown	0	1	0.05	0.22	430

Note: Std. dev. = standard deviation; No. = number; AADT = annual average daily traffic.

may also consider various other site characteristics, such as shoulder width, lane width, traffic control, and others.

Crashes are examples of count data and are typically modeled using a Poisson or negative binomial regression model. These belong to a category of models known as generalized linear models, or GLMs. The regression coefficients of GLMs and the standard error are estimated by maximizing the log-likelihood function of the observed parameters. Although the Poisson distribution provides a reasonable alternative to model crash data, the major drawback in using the model is that the Poisson distribution assumes the variance is equal to the mean. Crashes, by their nature of being rare and random occurrences are better modeled using a negative binomial framework that allows the variance to be greater than the mean. A negative binomial regression model was therefore used in this study to analyze the overdispersed crash data.

The natural log of the AADT was used as a predictor variable in the model and the length of the segment (mainline or ramp) was used as an offset variable, implying that crashes are explicitly assumed to increase proportionately with length. Several other predictor variables besides AADT can also be included in an SPF to better predict crashes. The coefficient of the predictor variable represents the impact of that variable on the predicted number of crashes.

This study includes an observation period of five years (2010–2014) and, as such, correlation is expected in annual crash counts at the same study location over the analysis period. This correlation violates the assumption of independently and identically distributed error terms and may potentially result in biased or inefficient parameter estimates. If the number of sites were small and the number of observations per site (i.e., years of data) large, a fixed effects model could be estimated wherein a unique site-specific constant would be estimated for each location. However, within the context of this study, the number of sites is large relative to the number of observations per site. Consequently, a random effects model is estimated, which allows the constant term of the negative binomial model to vary across sites. The mean and standard deviation of the constant term are estimated as additional model parameters. In this study, the random effects models were consistently shown to provide improved goodness-of-fit through a series of likelihood ratio tests.

Table 3. Aggregate Crash Data Trends for Interchange Ramps

Interchange configuration	Number of interchanges	Average annual crashes	Crashes per interchange per year
Diamond	278	2,304	8.3
Partial cloverleaf	96	1,179	12.3
Unconventional/hybrid	23	576	25.0
Trumpet	10	107	10.7
Full cloverleaf	8	171	21.4
Semi-directional	3	103	34.3
Three-leg directional	3	42	14.0
All directional	1	23	23.0
Single-point urban interchange (SPUI)	1	29	29.0

Table 4. Random Effects Negative Binomial Model for Interchange Mainline

Parameter	Estimate	Standard error	*t*-stat	*p*-value
Intercept	−7.39	0.37	−20.24	<0.01
log (AADT)	0.90	0.04	23.18	<0.01
Speed limit 70 mph (1 if yes; 0 otherwise)	−0.26	0.05	−4.66	<0.01
Speed limit <55 mph (1 if yes; 0 otherwise)	0.54	0.18	2.95	<0.01
Number of lanes greater than 4	0.22	0.10	2.29	0.02
Full cloverleaf (1 if yes; 0 otherwise)	0.33	0.14	2.34	0.02
Overdispersion parameter	0.04	0.01	5.88	<0.01

Results and Discussion

Exploratory Analysis

The data collected as a part of this study were used to conduct an exploratory crash analysis on the different interchange configurations to identify general trends at the interchange-configuration level. Table 3 provides a summary of the total crashes per interchange for each interchange configuration per year for existing interchange configurations in Iowa between 2010 and 2014.The diamond interchange, which is the most frequent interchange configuration in Iowa, also has the lowest total crashes per interchange configuration per year from the empirical study. Between the years of 2010 and 2014, an average of 8.3 crashes were observed to have occurred per diamond interchange per year and in the immediate vicinity of each diamond interchange located in Iowa. Partial cloverleaf (parclo) interchanges have a higher average number of total crashes occurring with 12.3 crashes occurring on average on each parclo in the state per year. Interchanges classified as semi-directional interchanges were observed to have the highest total crashes per interchange configuration per year in the study area. Single-point urban interchanges (SPUI) experienced the second highest number of crashes per interchange per year. However, it should be noted that the first SPUI in Iowa was opened to traffic in 2003 and the type is relatively new in the state. Moreover, the result is based on only one interchange. It should also be noted that there are very few directional interchanges in Iowa as well and the crash rate results shown for these interchange configuration (shown in Table 3) are based on a very small sample of those interchange configurations existing in Iowa.

A similar interchange exploratory analysis was carried out at the IIAs for select interchanges in Missouri. This research, however, considered only non-terminal-related crashes. The results from the study show that diamond interchanges experienced on average 5.5 total crashes per year and parclos experienced 6.0 total crashes per year. Full cloverleaf interchanges were also observed to experience 43.5 annual crashes per interchange (*14*). This is markedly higher than the results observed in this study. Because both states may have differing characteristics such as AADT and crash reporting procedures, it is expected that crash rates per interchange configuration may also vary. Subsequent to this exploratory analysis, a series of negative binomial models were estimated for both the mainline and ramp facilities based on the statewide interchange data.

Interchange Mainline Analysis

A series of negative binomial models were estimated in the statistics software. The analysis segment length was used as an offset variable in this model (i.e., it was

Table 5. Random Effects Negative Binomial Model for Interchange Ramps

Parameter	Estimate	Standard error	*t*-stat	*p*-value
Intercept	−5.02	0.22	−22.88	<0.01
log (AADT)	0.70	0.03	23.02	<0.01
Adv. Speed 10–35 mph (1 if yes; 0 otherwise)	−0.34	0.13	−2.67	<0.01
Signalized on-ramp (1 if yes; 0 otherwise)	0.35	0.12	2.90	<0.01
Signalized off-ramp (1 if yes; 0 otherwise)	0.84	0.10	8.13	<0.01
Overdispersion parameter	0.18	0.01	14.29	<0.01

assumed that crashes increase proportionally to length). AADT was included in log form and, as such, the parameter estimate for this variable indicates the elasticity of crashes with respect to volume. The other variables in each model are binary indicators, which distinguish differences among various speed limit categories and interchange configurations. Classical statistics assumes that observations are independent and identically distributed (IID). Within the context of this study, a 5-year analysis period was used and, as such, it is expected that crash counts would be correlated within the same locations (i.e., segments and ramps) over this period. Consequently, a random effect framework is utilized to account for this correlation. To this end, a site-specific random effect term is added that is unique to each analysis segment or ramp in the database. The resulting model is shown in Table 4. The random effects framework shows significant improvement as compared with a simpler, pooled model that does not consider the within-site correlation. Examining model goodness-of-fit statistics, both the log-likelihood and Akaike Information Criterion (AIC) show substantive improvements with the random effects model.

When examining the results from this model, as expected, crashes are found to increase roughly proportionally with respect to exposure (i.e., AADT). A 1% increase in volume is associated with a nearly elastic 0.96% increase in crashes along the interchange mainline. An observation of the speed limit parameter variables in the interchange mainline result shows a decrease in crashes as the speed limit on the mainline increases. A study carried out on interchanges in Florida showed comparable results with an increase in speed limit which was used as a continuous variable in this study resulting in a decrease in predicted crashes (*4*). It may appear counterintuitive that an increase in speed limit may result in fewer crashes, but this may be indicative of the more rural nature of the mainline locations that have higher speed limits. As the corridor is developed and the interchange location is more urbanized, it is assumed that the speed limit of such interchange mainlines would be reduced.

Differences among various interchange configurations in respect of crashes on the mainline were also investigated. The mainline of the full cloverleaf interchange configuration is associated with 34% more crashes when compared with the mainline of other interchange configurations.

Interchange Ramp Analysis

As with the mainline segments, a random effects negative binomial model was also estimated for the interchange ramps. The results from this model are presented in Table 5. In comparison with a simpler, pooled model, the effects of each factor are generally similar. The random effects model again shows significant improvements in both log-likelihood and AIC. Ramp length was treated as an offset variable in this model (i.e., the number of crashes increases/decreases in direct proportion to length). AADT is included in the model as a log-transformed continuous variable whereas other variables are discretized by introducing binary indicators to indicate the presence or absence of specific conditions.

Ramp advisory speeds from 10 to 35 mph are associated with fewer crashes on average. Many of the ramps with very low advisory speeds may also include additional supplementary signage, such as chevrons or other high-visibility traffic control devices, which may provide effective feedback to drivers, reducing speeds on entry to such ramps.

Figure 2 compares the observed crashes per mile per year on the interchange mainline and ramps with the crashes predicted from both the naïve pooled and random effects models. For the mainline random effects model, the prediction root mean square error (RMSE) is 0.78 whereas the RMSE for the naïve pooled model is 3.05. The random effects model for the ramps has an RMSE of 0.18 whereas the naïve pooled model has an RMSE of 1.02. Both metrics show the random effects models to provide significant improvements in goodness-of-fit as compared with the naïve pooled models. Similarly, Figure 2 provides a plot of observed crashes versus predicted crashes for the mainline, ramps, and the total IIA, and Figure 3 provides a series of cumulative residual plots for each facility type versus AADT. Collectively, these plots also show the clearly improved fit for the random effects model.

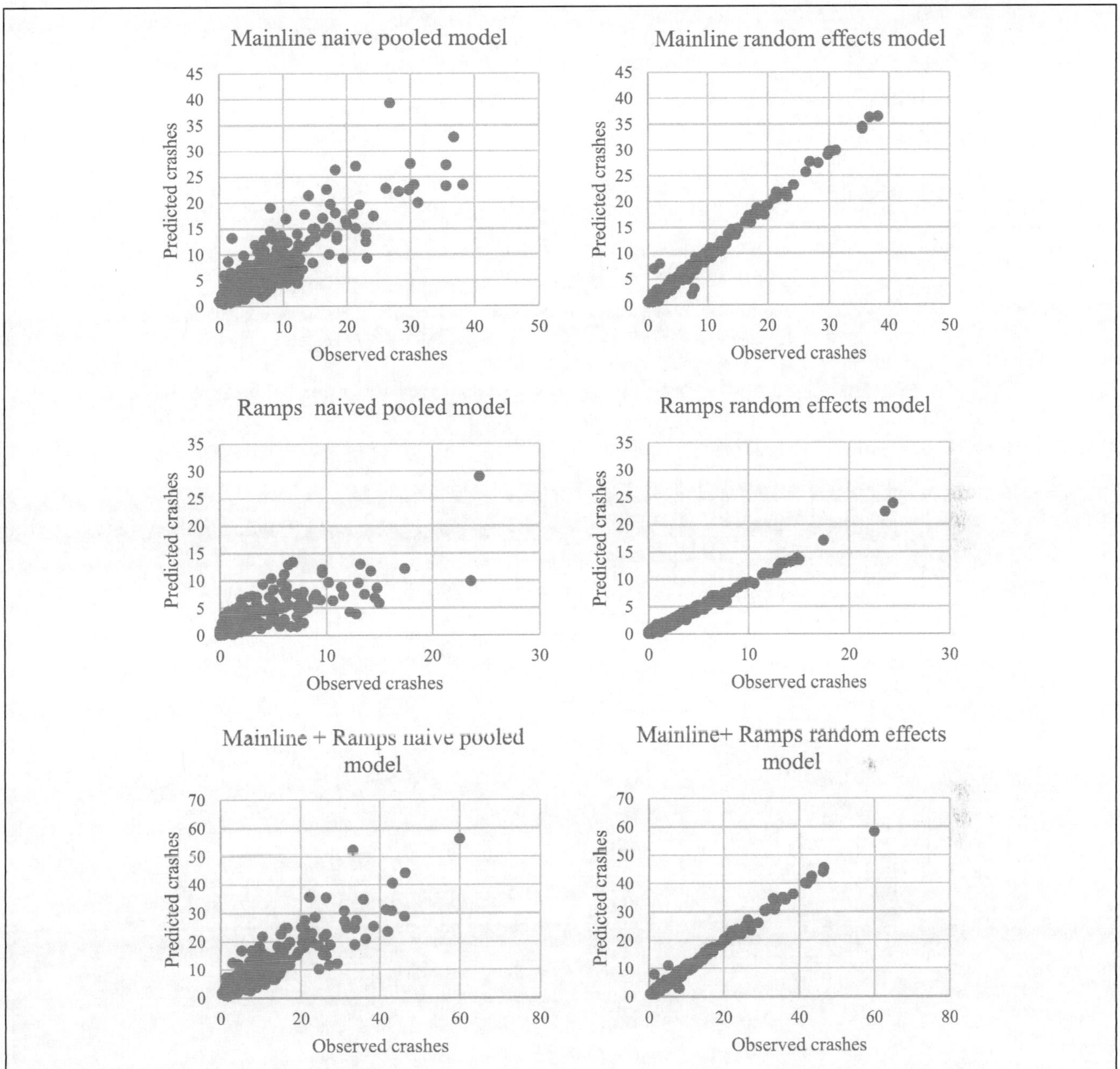

Figure 2. Observed crashes versus predicted crashes at IIA.

Table 6. Iowa-Specific AADT-Only SPFs for Freeway Mainline

Site subtype description	Intercept (Standard error)	*t*-stat (*p*-value)	AADT coefficient (Standard error)	*t*-stat (*p*-value)
Rural mainline—4 lanes	−6.34 (0.31)	20.45 (<0.01)	0.78 (0.03)	24.20 (<0.01)
Urban mainline—4 lanes	−7.65 (0.45)	17.00 (<0.01)	0.94 (0.05)	20.80 (<0.01)
Urban mainline—6 lanes	−3.25 (0.73)	4.45 (<0.01)	0.55 (0.07)	8.08 (<0.01)
Urban mainline—8 lanes	−9.54 (6.01)	1.59 (0.06)	1.13 (0.52)	2.16 (0.03)

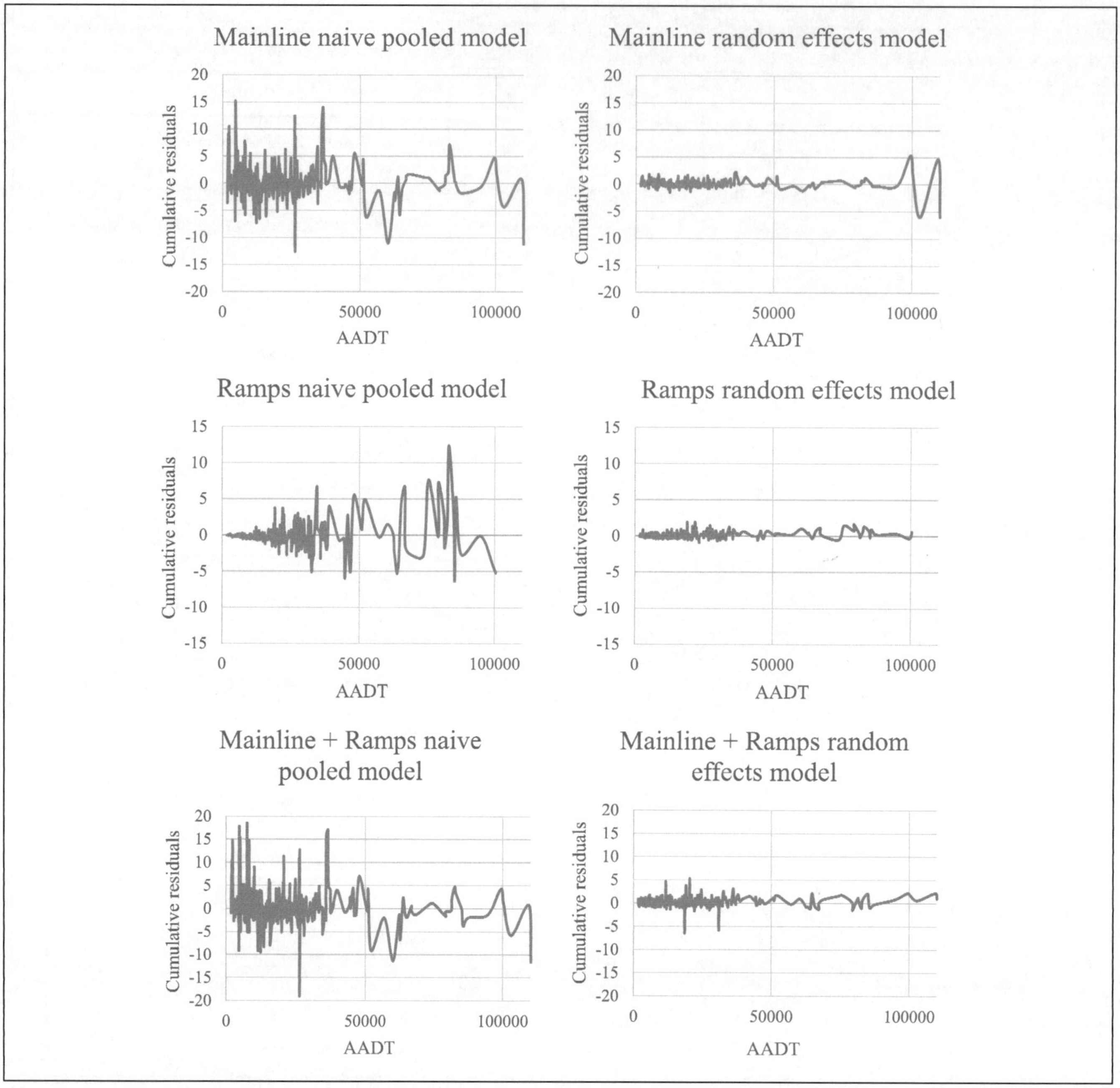

Figure 3. Cumulative residual plots for naïve pooled and random effects models.

Model Comparison with SafetyAnalyst and Florida-Specific SPFs

Table 6 shows simple SPFs developed for the interchange mainline using Iowa-specific data. In a study that compared Florida-specific simple SPFs at the interchange mainline to those developed for SafetyAnalyst, the results were observed to differ from the default national SafetyAnalyst SPFs (*13*). Figure 4 shows SPF graphs that compare Iowa-specific, Florida-specific, and SafetyAnalyst SPFs for the interchange mainlines. The Iowa-specific SPFs are observed to also differ from SafetyAnalyst SPFs, which provides additional support for the development of localized SPFs.

Conclusion

At the onset of the study, an exploratory crash analysis was conducted to examine safety trends for various interchange configurations. This analysis resulted in general crash trends for various common interchange

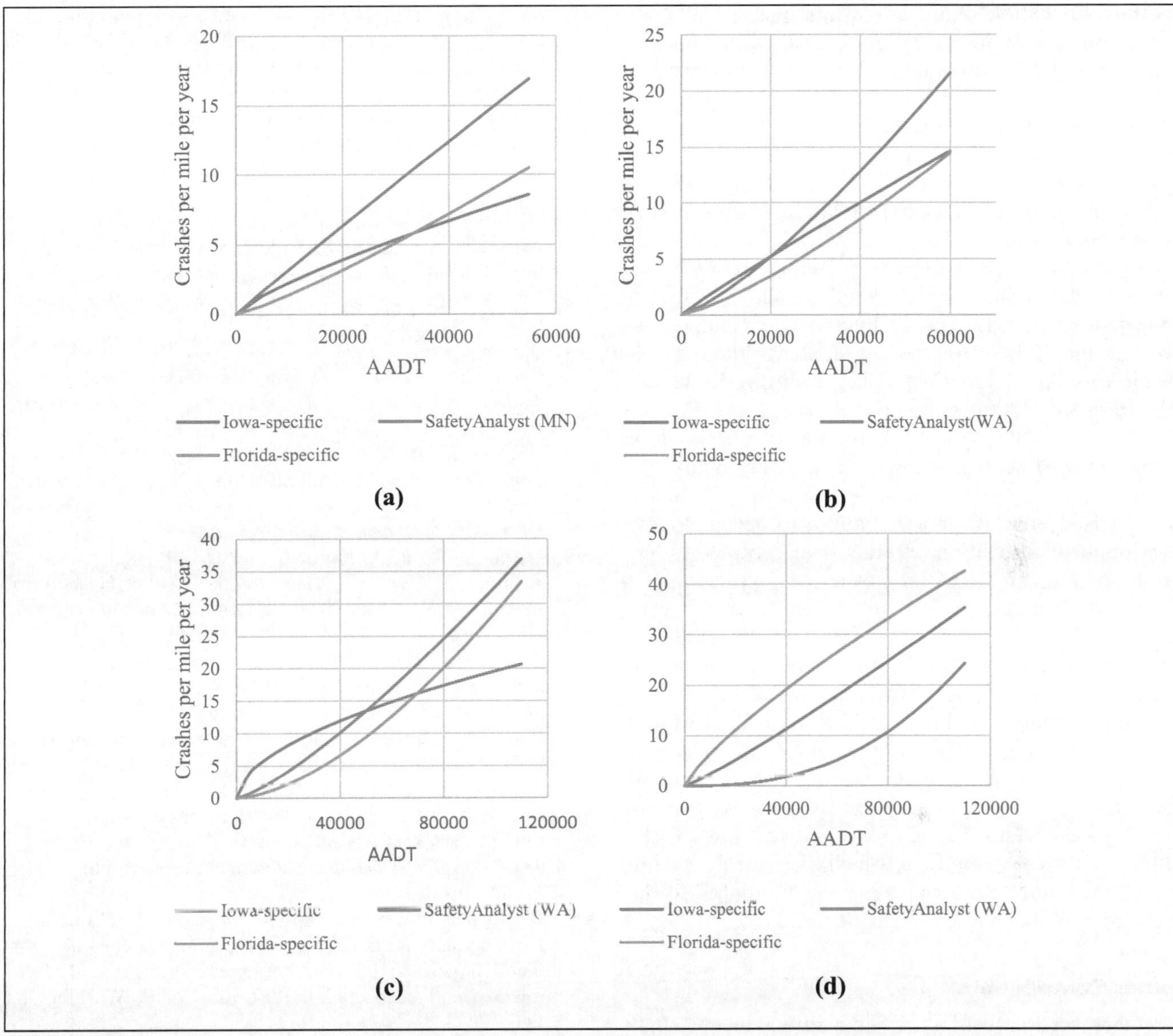

Figure 4. Comparison of interchange mainline SPFs between Iowa and other states: (*a*) total crashes rural mainline—4 lanes; (*b*) total crashes urban mainline—4 lanes; (*c*) total crashes urban mainline—6 lanes; (*d*) total crashes urban mainline—8 lanes. Florida and SafetyAnalyst SPFs are for 8+ lanes.

configurations. These high-level trends provide general insights as to the safety performance of various interchange alternatives. They also motivate the need for more detailed investigations to distinguish the reasons for these differences in safety performance.

The diamond interchange, which is also the most common interchange configuration in Iowa, showed the lowest frequency of crashes per interchange. This may be reflective of the simple nature of the diamond interchange, which is applicable in most rural and suburban settings. In contrast, unconventional/hybrid interchanges, as well as the only SPUI in the state of Iowa, experienced more crashes in comparison with other interchange configurations. This provides statistical support to longstanding guidance, which recommends simplicity and consistency of interchange design across a jurisdiction (*15*).

A primary contribution of this research was the development of Iowa-specific SPFs for mainline interstates and ramp facilities. Interestingly, those mainline facilities with higher speed limits experienced fewer crashes on average than those with lower posted speed limits. It is important to acknowledge that this finding may simply be reflective of the way speed limits are generally reduced in more urbanized areas because of increased ramp density, space constraints, and other factor that make such roadways susceptible to higher crash frequencies. In contrast, crashes were significantly reduced on interchange ramps with the lowest advisory speeds. Consequently, additional efforts are warranted to better understand the speed–safety relationship. Signalized ramp terminals also showed elevated crash risk, particularly at the off-ramps if the terminal was signalized. Multi-vehicle rear-end crashes contribute a substantial amount to crashes occurring at signals (*16*).

The SPFs developed were also compared with those from SafetyAnalyst, as well as results of similar research in the state of Florida. For the interchange mainline, the Iowa-specific SPFs were generally similar to those developed in Florida, and markedly different from the default SPFs from SafetyAnalyst.

From a methodological standpoint, the random effects framework was shown to provide significant improvement in goodness-of-fit as compared with naïve pooled (fixed effects) models. Failure to accommodate the within-site correlation resulting from repeated observations of the same locations over the 5-year analysis period would otherwise lead to biased parameter estimates. The random effects are able to accommodate site-specific heterogeneity, resulting in a significant reduction in the overdispersion parameter (from 0.18 to 0.04 for the mainline models and from 0.92 to 0.04 for the ramp models). An important implication of this result is that the overdispersion parameter resulting from a random effects model is different in nature from that of a naïve pooled model. This has implications for use of the Empirical Bayes method, in which the overdispersion parameter is used to determine the weight applied to the predicted and observed crash frequencies.

Author Contributions

The authors confirm contribution to the paper as follows: study conception and design: PS, AA; data collection: AA; analysis and interpretation of results: PS, AA; draft manuscript preparation: PS, AA. All authors reviewed the results and approved the final version of the manuscript.

References

1. Bonneson, J. A., S. R. Geedipally, M. P. Pratt, and D. Lord. *NCHRP Program Project 17–45: Safety Prediction Methodology and Analysis Tool for Freeways and Interchanges*. Transportation Research Board of the National Academies, Washington, D.C., 2012.
2. Torbic, D., D. Harwood, D. Gilmore, K. Richard, and J. Bared. Safety Analysis of Interchanges. *Transportation Research Record: Journal of the Transportation Research Board*, 2009. 2091: 39–47.
3. Kiattikomol, V. *Freeway Crash Prediction Models for Long-Range Urban Transportation Planning*. Doctoral dissertation. The University of Tennessee, Knoxville, TN, 2005. https://trace.tennessee.edu/utk_graddiss/2161/.
4. Kobelo, D. Interchange Crash Analysis by Type on Florida Highways. *International Refereed Journal of Engineering and Science*, Vol. 2, No. 10, 2013, pp. 31–40.
5. Torbic, D. J., D. W. Harwood, D. K. Gilmore, and K. R. Richard. *Interchange Safety Analysis Tool (ISAT): User Manual*. U.S. Department of Transportation, Federal Highway Administration, Washington, D.C., 2007.
6. American Association of State Highway and Transportation Officials. *Highway Capacity Manual 2010*, 1st ed. American Association of State Highway and Transportation Officials, Washington, D.C., 2010.
7. Papadimitriou, E., and A. Theofilatos. Meta-Analysis of Crash Risk Factors in Freeway Entrance and Exit Areas. *Journal of Transportation Engineering, Part A: Systems*, Vol. 143, No. 10, 2017.
8. Riniker, K. A. The Folded Interchange: An Unconventional Design for the Reconstruction of Cloverleafs. *Proc., ITE 2009 Technical Conference and Exhibit, Transportation Operations in Action*, Phoenix, AZ, 2009.
9. Green, E. R., R. A. Kenneth, and J. G. Pigman. *Research Report: Kennedy Interchange Crash Study*. University of Kentucky, Kentucky Transportation Center, Lexington, KY, 2010.
10. Gan, A., K. Haleem, P. Alluri, J. Lu, T. Wang, M. Ma, and C. Diaz. *Preparing Florida for Deployment of SafetyAnalyst for All Roads*. Lehman Center for Transportation Research Florida International University in Cooperation with Research Center, State of Florida Department of Transportation (FDOT), Miami, FL, 2012.
11. Moon, J., and J. Hummer. Development of Safety Prediction Models for Influence Areas of Ramps in Freeways. *Journal of Transportation Safety and Security*, Vol. 1, No. 1, 2009, pp. 1–17.
12. Bornheimer, C., S. Schrock, M. Wang, and H. Lubliner. Developing a Regional Safety Performance Function for Rural Two-lane Highways. Presented at 91st Annual Meeting of the Transportation Research Board, Washington, D.C., 2012.
13. Lu, J., K. Haleem, A. Gan, and P. Alluri. Safety Performance Functions for Florida's Freeways. *Proc., Second Transportation & Development Conference*, Orlando, FL, 2014, pp. 321–331.
14. Zhang, M. *Empirical Study of Freeway Interchange Crash Characteristics and Influence Areas*. Master's Thesis, University of Missouri, Columbia, MO, 2016.
15. Mulinazzi, T. E. *Final Report: Guidelines for the Selection of an Interchange Configuration*. Joint Highway Research Project, Purdue University, West Lafayette, IN, 1973.
16. Yan, X., E. Radwan, and M. Abdel-Aty. Characteristics of Rear-End Accidents at Signalized Intersections Using Multiple Logistic Regression Model. *Accident Analysis and Prevention Journal*, Vol. 37, No. 6, 2005, pp. 983–995.

The Standing Committee on Highway Safety Performance (ANB25) peer-reviewed this paper (18-00011).

TRR
JOURNAL OF THE TRANSPORTATION RESEARCH BOARD

Article

Transportation Research Record
2018, Vol. 2672(30) 131–141
© National Academy of Sciences:
Transportation Research Board 2018
Article reuse guidelines:
sagepub.com/journals-permissions
DOI: 10.1177/0361198118793498
journals.sagepub.com/home/trr
SAGE

Evaluating the Impact of Rumble Strips on Fatal and Injury Freeway Crashes

Bahar Dadashova[1], Lingtao Wu[2], and Karen Dixon[1]

Abstract

Rumble strips are known to be one of the most cost-effective treatments for preventing roadway departure crashes. However, in recent years some studies have found controversial results indicating that rumble strips may in fact increase the number of more severe crashes. Although these effects are estimated to be very small, highway safety agencies deploy these treatments with an expectation that they will reduce all crash types. In this paper, the authors have conducted a statistical evaluation to determine the impact of rumble strip presence on fatal and injury crashes at freeway locations. For this purpose, the authors acquired an existing database used for one of the aforementioned studies and evaluated the variables using alternative assessment methods. The results of the current study suggest that rumble strips do in fact improve the safety outcomes in rural freeways. These findings are observed to also apply to urban freeways, but the effects are not statistically significant for the study database.

Rumble strips are known to be one of the most cost-effective treatments for preventing roadway departure crashes (often referred to as run-off-the-road (ROR) crashes) and driver fatigue-related crashes. Rumble strips are particularly known to be effective at reducing crashes at horizontal curve locations (*1–3*). Studies have revealed that rumble strips may reduce ROR crashes by 10% to 79% (*1, 3–7*).

Nevertheless, studies in recent years have reported controversial results which indicate that fatal crashes may in fact increase as the result of the rumble strip installation. As an example, this observation is documented in the final report for National Cooperative Highway Research Program (NCHRP) 17-45 (*8*). In the NCHRP study, the authors developed crash modification factors for fatal and injury crashes on tangent and curve segments, and found that on the freeway tangents the rumble strips could potentially contribute to a 19% decrease in a combined value that includes fatal and injury crashes, but when the rumble strip is evaluated independently for single-vehicle fatal or single vehicle incapacitating injury crashes, the number of crashes can be expected to increase. A few studies have also reported similar results. Torbic et al. (*1*) reported 2%–40% increase in crashes on arterials and freeways with shoulder rumble strips. Hu and Donnell (*9*) found that the presence of shoulder rumble strips increases the possibility of fatal and incapacitating injury crashes. It is worth mentioning that all the three studies utilized regression model methods to estimate the safety effects of rumble strips. Recently, researchers have argued that the conventional regression model methods may not capture the causal and effect relationship between countermeasures and traffic crashes (*10*).

This paper reports on a study that has been developed with the goal of implementing robust causal inference models to explore the relationships between rumble strips and crash occurrences. For this purpose, the authors have acquired and used the original NCHRP 17-45 data. To conduct the data analysis, the authors have disaggregated the data into rural and urban segments, and have implemented exploratory and causal inference analysis. Incidence rate ratio was used to conduct the exploratory analysis of the observed data. The authors used propensity score matching (PSM) to conduct the causal inference analysis. In recent years, PSM has gained significant attention in the transportation safety research area as a tool that can help account for site selection bias.

Data Description

As indicated earlier, authors had acquired the original data used in the NCHRP 17-45 report to conduct the data analysis in the current study. In the aforementioned

[1]Texas A&M Transportation Institute, College Station, TX
[2]Texas A&M Transportation Institute, Texas A&M University System, College Station, TX

Corresponding Author:
Address correspondence to Bahar Dadashova: b-dadashova@tti.tamu.edu

study, researchers had used the freeway data collected from the following three states:

- California (2005–2007)
- Washington (2005–2007)
- Maine (2006–2008)

The data were obtained from the Highway Safety Information System (HSIS) database. The HSIS data includes the crash and roadway geometric design characteristics. However, the HSIS database does not include rumble strip information. Consequently, the research team of NCHRP 17-45 had acquired this information as a supplement to the data.

Rumble Strips

Most of the rumble strips included in NCHRP 17-45 were milled into the pavement (Figure 1*a*). For that study, the researchers measured the length of the rumble strips for both the inside (1 & 4) and the outside (2 & 3) shoulders for both directions of travel for a given segment. Both in the original and current study the authors had used the freeway segments defined by the HSIS database. The researchers then calculated the proportion of rumble strips to the segment length (PRS) as shown in Equation 1:

$$\text{PRS} = \frac{\text{Length of Rumble Strips}}{\text{Length of Segment}} \tag{1}$$

The researchers calculated the PRS for all four locations on each segment. To obtain a single PRS per location, they then averaged the four measures. In Figure 1*b* the PRS distribution for all segments is graphically depicted. The figure shows that approximately 58.2% of the segments did not have rumble strips. For 0.5% of all segments, the PRS was equal to 0.1. Similarly, there were 0.3% of all segments where the PRS value was equal to 0.2. Finally, 23.6% of segments had fully installed rumble strips or a PRS value equal to 1.0.

When examined separately, the authors observed that the distribution of the PRS for rural vs urban freeway

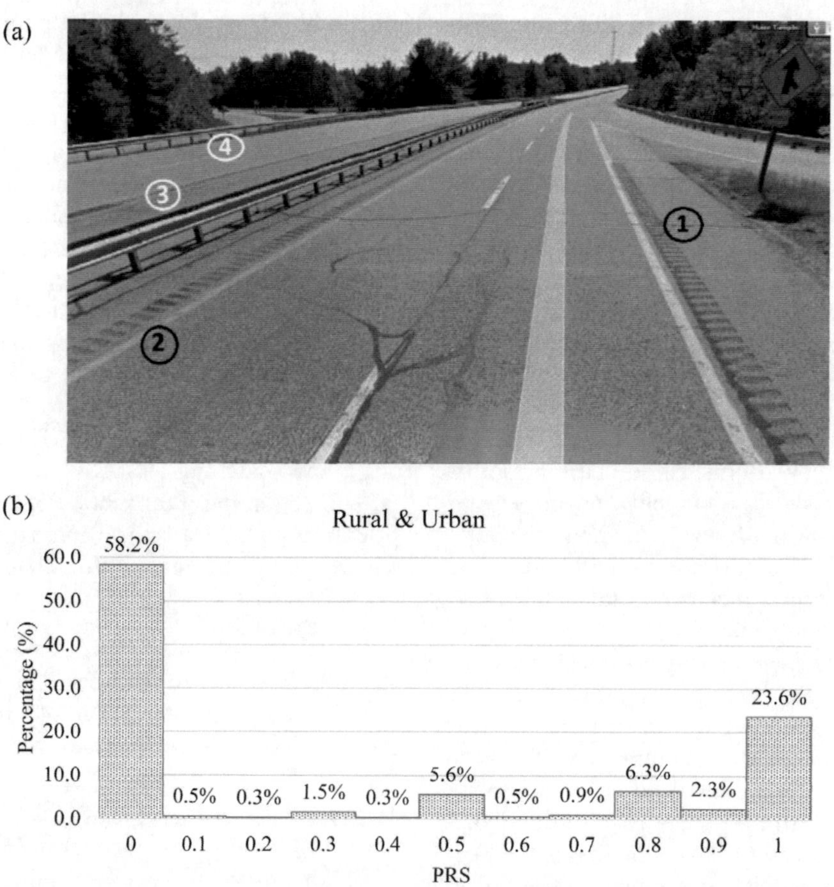

Figure 1. Proportion of rumble strips.

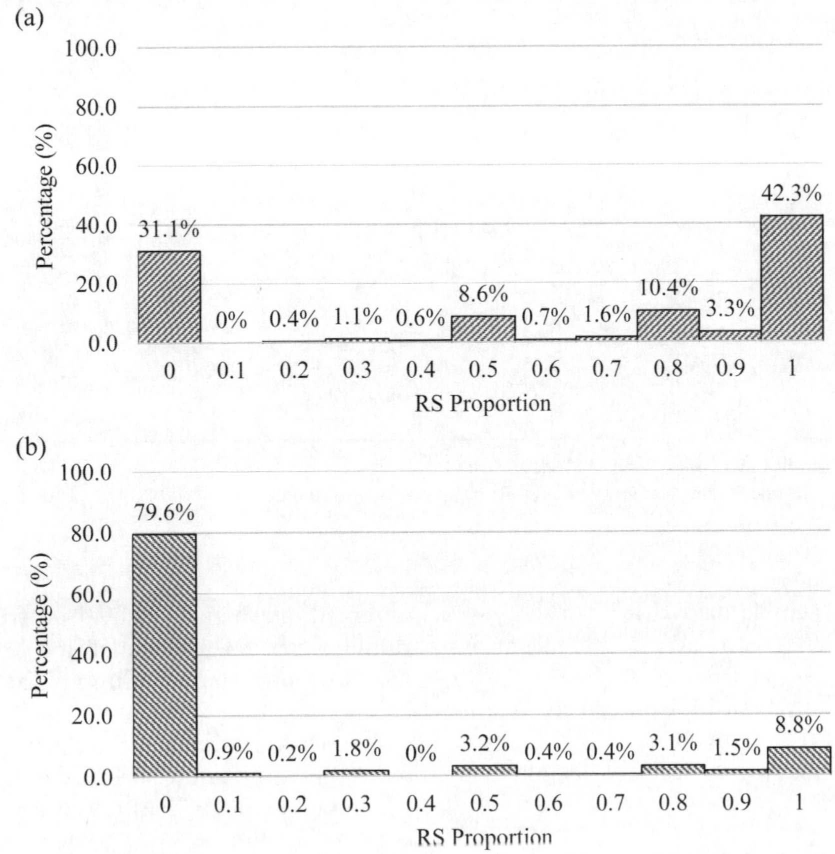

Figure 2. Percentage of (*a*) rural, and (*b*) urban segments with rumble strips.

Table 1. Treatment Status of Segments Based on the PRS

Rumble strips	Status	No. of rural segments	No. of urban segments	Treatment status
PRS = 0	Untreated	263	839	0
PRS<0.75	Treated	15	31	1
0.75≤PRS<1	Treated	234	102	1
PRS = 1	Treated	315	81	1
Total		827	1,053	

Note: PRS = proportion of rumble strips.

segments varied significantly (Figure 2). On rural freeways, 42.3% of segments had rumble strips fully installed, while only 8.8% of the urban freeways had rumble strips applied along all shoulders. A common concern about constructing rumble strips on urban freeways is that the noise generated from the resulting "rumble" is not compatible with the context of the urban environment. It is likely, therefore, that rumble strips are applied more often on rural freeways compared to urban freeways. Therefore, the authors conducted the data analysis separately for rural and urban freeway segments.

In this study, authors assume that the treatment variable (rumble strip) is a dummy variable which takes value of either zero or one indicating the status of treatment (i.e. 1 = Treated and 0 = Untreated). For this purpose, authors divided the segments into four samples based on the treatment status of segments as shown in Table 1:

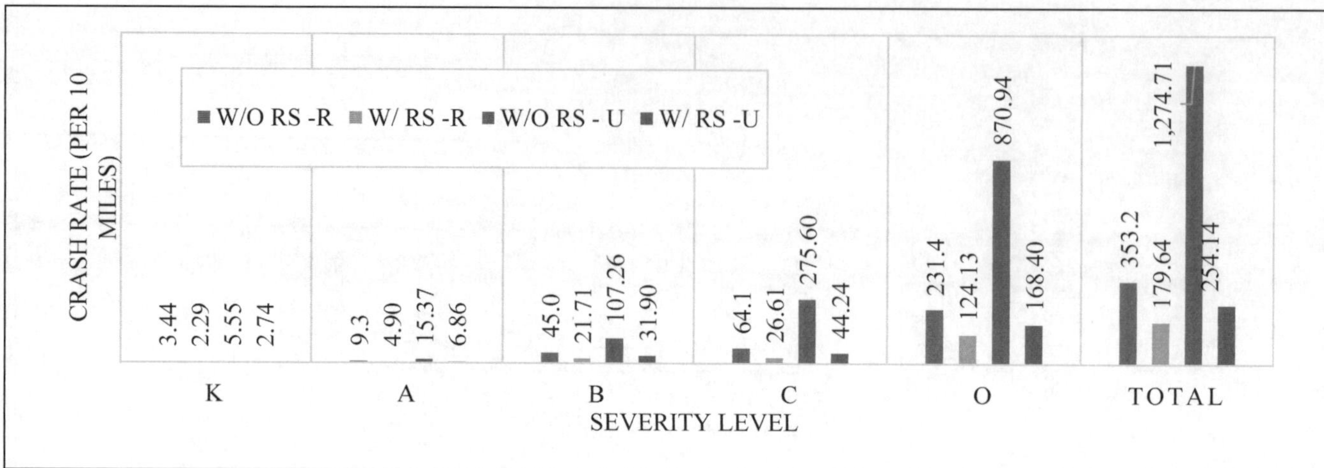

Figure 3. Crash rate per 10 miles vs rumble strip presence.
Note: W/O RS-R = Rural segments without rumble strips; W/RS-R = Rural segments with rumble strips; W/O RS-U = Urban segments without rumble strips; W/RS-U = Urban segments with rumble strips.

- Segments with no rumble strips, PRS = 0.
- Segments where less than 75% of the segment length had rumble strips, PRS < 0.75.
- Segments where more than 75% of segment length had rumble strips, 0.75 ≤ PRS < 1.
- Segments where the entire segment length had rumble strips, PRS = 1.

Since the number of segments belonging to the second category (i.e. PRS < 0.75) were small (15 rural and 31 urban segments) these segments were included in the study as "Treated" sites. As a result there were 263 Untreated and 564 Treated segments on rural freeways, and 839 Untreated and 214 Treated segments on urban freeways.

Crash Data

Because the counterintuitive findings from previous studies noted that fatal and incapacitating injury crashes appeared to increase at locations with rumble strips, the authors evaluated crash data for a variety of injury and crash types including:

- Single-vehicle (SV) crashes.
- Multiple-vehicle (MV) crashes.
- Crashes located at ramp exits and entrances.
- Fatal (K) crashes.
- Incapacitating injury (A) crashes.
- Non-incapacitating injury (B) crashes.
- Possible injury (C) crashes.
- Property damage only (O) crashes.

The crash rate per 10 miles for both rural and urban segments is depicted in Figure 3. As can be observed, crash rates located on freeway segments with rumble strips (W/RS) are always less than those on segments without rumble strips (W/O) for all crash severity types.

K and A crashes are extremely rare events and often quite similar in nature. Because of the limited sample size for these two individual crashes, the authors elected to conduct a joint analysis of K and A crashes. In this study, therefore, the authors have conducted the data analysis for rumble strip effectiveness at rural and urban freeway segments for the following crashes:

- Multiple vehicle KA crashes (MV-KA).
- Multiple vehicle B crashes (MV-B).
- Single vehicle KA crashes (SV-KA).
- Single vehicle B crashes (SV-B).

Potential Confounding Factors

In this study, the authors have used several candidate treatment covariates to assess the effectiveness of rumble strips in the presence of additional roadway conditions that could be correlated to rumble strip effectiveness. The list of these variables and their descriptive statistics are described in Table 2. The K-factor refers to the proportion of vehicles at the peak hours. A curve segment refers to the segment with at most three horizontal curves. Therefore, each curve could have up to three radii as shown in the table. The authors have also used the ramp exit/entrance dummy variable showing whether the segment had a merging and/or a diverging lane. At many of these entry or exit ramp locations, rumble strips are not continuous due to the merging and diverging of the additional travel lanes.

Table 2. Descriptive Statistics of Variables

Variables	Min	Max	Mean	S.D.
Response (No. of crashes)				
Rural MV-KA	0	3	0.11	0.37
Rural SV-KA	0	6	0.2	0.55
Rural MV-B	0	10	0.35	0.88
Rural SV-B	0	20	0.7	1.37
Urban MV-KA	0	4	0.14	0.46
Urban SV-KA	0	4	0.14	0.44
Urban MV-B	0	15	0.69	1.43
Urban SV-B	0	13	0.68	1.31
Treatment covariates				
Rumble Strips	0 - not treated (control), 1 - treated			
Median width (ft)	2	429	41.79	7.04
Lane width (ft)	10	15	12.15	0.71
Shoulder width (ft)	2	20	8.13	1.39
K-factor	0	0.93	0.23	0.56
Average AADT	8,173	307,666	78,880.89	241.94
Number of lanes	4	10	5.72	1.58
Proportion of curve	0	1	0.26	0.61
Proportion of barrier	0	0.95	0.17	0.45
Curve #1 Radius (ft)	0	20,000	2,107.84	57.21
Curve #2 Radius (ft)	0	20,000	255.35	37.02
Curve #3 Radius (ft)	0	10,000	32.51	21.31
California (CA)	Dummy variables indicating the location of the segment			
Washington (WA)				
Maine (MA)				
Curve/Tangent	0 - tangent, 1 - curve			
Ramp exit/entrance	0 - no ramp exit/ entrance on the segment, 1 - at least one ramp exit/entrance			

Note: MV-KA = multiple-vehicle fatal and injury crashes; SV-KA = single-vehicle fatal and injury crashes; MV-B = multiple-vehicle non-incapacitating injury crashes; SV-B = single vehicle non-incapacitating injury crashes.

Methodological Approach

The data available for this analysis provides cross-sectional information. This means that a statistical before–after study is not feasible. The cross-sectional data could potentially introduce a selection bias if the maintaining agency has installed the rumble strips only at high crash frequency zones. In many cases, rumble strip applications are systemic and so are not affected by this issue. To address any potential selection bias issues, the authors have conducted exploratory and causal data analysis to evaluate the impact of rumble strip presence on the crash severity. Namely, the authors used an incidence rate ratio to conduct the exploratory analysis of KA crashes and a propensity score model to conduct the causal inference analysis of both KA and B crashes.

Incidence Rate Ratio

The incidence rate ratio (IRR) is an exploratory data analysis tool. The IRR measures the strength of the association between a treatment method and the event. IRR is frequently used in epidemiological studies to explore the impact of the treatment method on the disease and can be interpreted as the odds ratio if the disease is a rare event. Given that the KA crashes are rare events, the IRR estimated in this study can therefore be interpreted as an odds ratio.

The IRR is calculated as the ratio of two incidence rates (or risk rates),

$$\mathrm{IRR}_{\mathrm{KA}} = \frac{P(\mathrm{KA} = 1|\mathrm{RS} = 1)}{P(\mathrm{KA} = 1|\mathrm{RS} = 0)} \tag{2}$$

where

$P(\mathrm{KA} = 1|\mathrm{RS} = 1)$ is the incidence rate or probability of a KA crash occurring given that there are rumble strips, and $P(\mathrm{KA} = 1|\mathrm{RS} = 0)$ is the incidence rate or probability of a KA crash occurring given that there are no rumble strips.

The IRR then indicates how prone the segment is to KA crashes based on the presence of a rumble strip and is interpreted as

$$\text{Rumble Strips are associated with} \begin{cases} \text{More KA crashes, if } \mathrm{IRR}_{\mathrm{KA}} > 1 \\ \text{Less KA crashes, if } \mathrm{IRR}_{\mathrm{KA}} < 1 \end{cases} \tag{3}$$

Using the Bayes theorem, the incidence rate of a KA crash on a segment with rumble strips can be calculated as

$$P(\mathrm{KA}=1|\mathrm{RS}=1) = \frac{P(\mathrm{RS}=1|\mathrm{KA}=1)\times P(\mathrm{KA}=1)}{P(\mathrm{RS}=1)} = \frac{P(\mathrm{RS}=1|\mathrm{KA}=1)\times P(\mathrm{KA}=1)}{P(\mathrm{RS}=1|\mathrm{KA}=1)\times P(\mathrm{KA}=1)+P(\mathrm{RS}=1|\mathrm{KA}=0)\times P(\mathrm{KA}=0)} \tag{4}$$

Similarly, the incidence rate of a KA crash on a segment without rumble strips (called the control group from this point forward) will be equal to

$$P(\mathrm{KA}=1|\mathrm{RS}=0) = \frac{P(\mathrm{RS}=0|\mathrm{KA}=1)\times P(\mathrm{KA}=1)}{P(\mathrm{RS}=0)} = \frac{P(\mathrm{RS}=1|\mathrm{KA}=1)\times P(\mathrm{KA}=1)}{P(\mathrm{RS}=0|\mathrm{KA}=1)\times P(\mathrm{KA}=1)+P(\mathrm{RS}=0|\mathrm{KA}=0)\times P(\mathrm{KA}=0)} \tag{5}$$

Propensity Score Matching

The predictive methods described in the Highway Safety Manual (HSM) assume that the "counterfactual" crash frequency, i.e. the crash frequency that would have been observed if the countermeasure had not been applied, can be estimated by applying a base safety performance function (SPF) model (*11*). In causal inference models, the effect of the countermeasure is estimated by comparing the counterfactual crash frequency to the observed crash frequency.

Causal inference models are based on matching a set of "potential observations" for each population of observed crash frequency to unobserved counterfactual crash frequency (i.e. the control group) (*12*). The effect of a treatment is then studied by employing non-randomized data selected from potential observations. Casual inference model uses the *propensity score* to mimic the random selection method. A propensity score represents a conditional probability of a facility receiving a countermeasure given the covariates and the outcomes. It shows the relationship between treatment status (1 - Treated; 0 - Control) and covariates (i.e. variables that completely or partially account for the apparent association between an outcome and risk factor).

Propensity score can be estimated using several parametric and non-parametric tools. In safety analysis literature, discrete choice logistic models are most commonly applied (*13–15*). In this study, the authors have used both random forests and discrete choice (logistic regression) models to estimate the propensity score. Figure 4 depicts the PSM framework used in this study.

Data Analysis Results

Exploratory Data Analysis

To assess the differences between the segments with and without rumble strips, and to observe their relationship to the KA crash occurrence, the authors conducted an exploratory data analysis. The IRR can be calculated for four samples (rural and urban freeways with and without horizontal curvature) as shown in Table 3.

The IRRs for the four study samples are shown in Table 4. For rural curve segments, the IRR is equal to 0.72, therefore indicating that in the observed data, rural freeway segments with horizontal curvature and rumble strips had experienced 28% (i.e. 100%–72%) less crashes than the freeway curve segments without rumble strips.

As can be observed, curve segments with rumble strips for both rural and urban locations experienced fewer KA crashes than the ones observed at tangent locations. On the other hand, the tangents with rumble strips had experienced more KA crashes. These results are counter-intuitive. One reason is because the segments labelled as curves include tangent sections as well (a curve can have three curve sections and the tangents connecting the curves), thus the tangent segments included in this data might not be the true representative of existing tangents in the data. In order to avoid biased estimations such as these, researchers used rural and urban segments and did

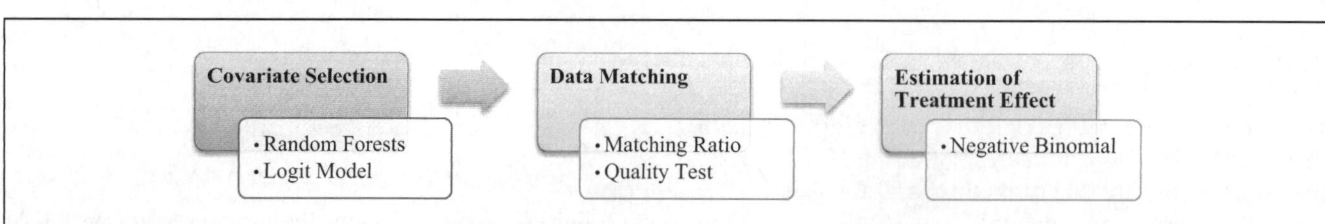

Figure 4. PSM framework.

Table 3. Data Sample Size for Four Datasets

Area type	Rumble strip present	Number of segments with rumble strips		Total number of segments
		Curves	Tangents	
Rural	Yes	206	358	564
	No	110	153	263
Urban	Yes	109	105	214
	No	369	470	839
Total		794	1,086	1,880

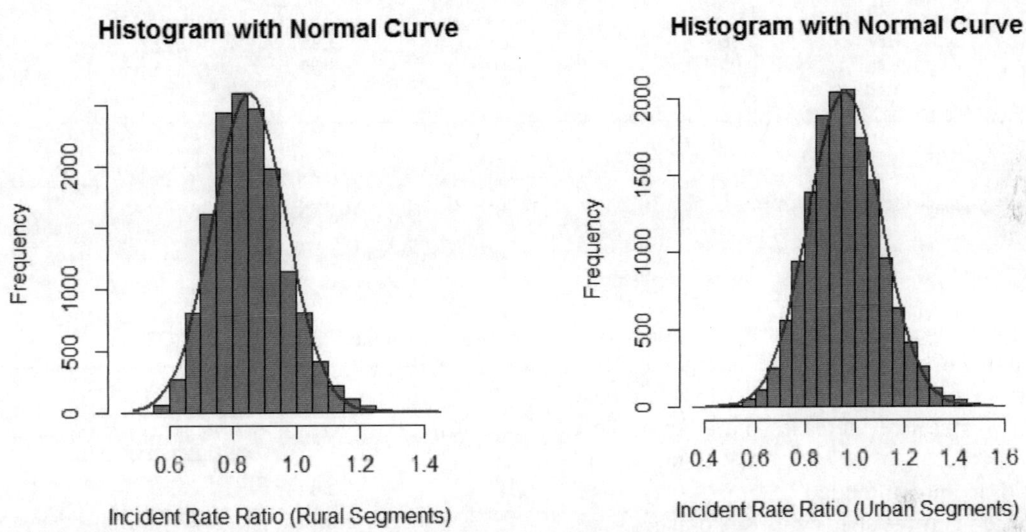

Area Type	Min	Max	Median	Mean	Std. Error	CI (95%)	
						LL	UL
Rural	0.48	1.44	0.84	0.85	0.11	0.62	1.07
Urban	0.42	1.61	0.96	0.96	0.14	0.68	1.25

Figure 5. Distribution of population IRR estimates.

Table 4. IRR for Four Samples

Area type	IRR_{KA}		
	Curves	Tangents	All
Rural	0.72	1.22	0.84
Urban	0.83	1.09	0.96

not differentiate between curve and tangent segments for the rest of the data analysis.

Bootstrap Sampling. In order to obtain the IRR of the population rather than the samples, the authors used the bootstrap method. Bootstrapping is a resampling method with replacement. Using the bootstrap method, the authors simulated 15,000 new samples and calculated the IRR for rural and urban freeway segments separately. The descriptive statistics and the distribution the IRR are shown in Figure 5.

The mean and median values of the IRR indicate that overall, segments with rumble strips on both rural and urban areas have lower probability of experiencing KA crashes; rural and urban segments with rumble strips would experience 15% (1 − 0.85 = 0.15) and 4% (1 − 0.96 = 0.04) less KA crashes compared with the segments with no rumble strips.

Confidence intervals show the lower limit and upper limits of these effects. As the presence of rumble strips increases on the rural freeways, the changes in KA

Table 5. Covariates Correlated with the Rumble Strip Presence on Freeway Segments

Covariate	Rural freeway segments				Urban freeway segments			
	Variable importance[a]	*p* value	Logit Estimator[b]	*p* value	Variable importance	*p* value	Logit estimator	*p* value
Median width[c]	76.17	<0.01	0.03	<0.001	69.76	<0.01	0.01	<0.001
K-factor[c]	40.87	<0.01	−2.29	<0.001	60.78	<0.01	−3.58	<0.001
CA[c]	19.01	<0.01	−3.52	<0.001	8.5	<0.01	−1.94	<0.001
No. of lanes[c]	17.18	<0.01	−0.14	<0.001	18.75	<0.01	−0.22	<0.001
WA[c]	15.43	<0.01	−2.98	<0.001	13.37	<0.01	−3	<0.001
Lane width	53.85	0.91			39.06	1		
Proportion of barrier	28.18	1			34.06	1		
Curve #1 Radius	22.83	0.87			29.66	0.5		
Proportion of curve	15.93	1			15.58	0.5		
Ramp exit/entrance	5.62	1			4.47	1		
Curve #2 Radius	4.27	0.86			3.29	0.92		
Curve dummy	2.38	0.87			2.37	0.65		
Curve #3 Radius	0.36	0.67			0.82	0.5		

[a]Measured using random forests.
[b]Estimates of logistic regression.
[c]Selected covariates for matching.

crashes could vary from a 48% decrease to a potential 7% increase. On the urban freeway segments, the KA crashes could vary from a 42% decrease to 25% increase as the rumble strip presence increases. Note that these findings are the results of exploratory analysis of observed data and are not intended to imply any causal relationships between rumble strips and KA crashes.

Propensity Score Matching

Covariate selection. To identify the characteristics of the sites with and without rumble strips, authors conduct the random forest and logistic regression analysis. The random forests classification tool is used to select most important variables that are significantly correlated with the rumble strip presence on freeways. The variables referred to as median width, K-factor, number of lanes, and locations in California and Washington were found to be very important for the rumble strip presence on a freeway segment. To cross-validate this assumption, the authors used the logistic model to estimate the impact of these variables on the rumble strips. The results show that all five variables are significantly correlated with the rumble strip presence on both rural and urban freeways (Table 5).

Data Matching. Following identification of the significant covariates, the authors matched the treated segments and control segments using the selected variables and a matching ratio. It is assumed that the ratio used to match the treated observations with the control group observations should be higher than 1:2, i.e. each treated segment has to be matched with at least two control segments (*15*) . Nevertheless, given that the number of treated rural segments is higher than the number of control segments in this pre-existing database, the authors used a 1:1 ratio to match the rural segments and a 1:2 ratio to match the urban segments. The sample sizes of the matched datasets are shown in Table 6.

Table 6. Sample Size for Matched Data

Sample size	Rural		Urban	
	Treated	Control	Treated	Control
All	564	263	214	839
Matched	134	194	157	164
Unmatched	430	69	57	675
Removed	0	0	0	0

Matching Quality Test. Prior to the count data modeling, the quality of the matching data needs to be tested and validated. To validate the quality of the matched data the propensity score distributions are inspected (Figure 6*a*). It is expected that when overlaid, the propensity score distributions of treated and control groups will overlap. The overlapping area is referred to as common support. The observations falling beyond the area need to be discarded. As can be observed for both the rural and urban samples, all of the matched observations fall within the same area.

The second quality control test is data balancing. It is expected that the covariate means in both treatment and control groups are very similar. To test this claim, the

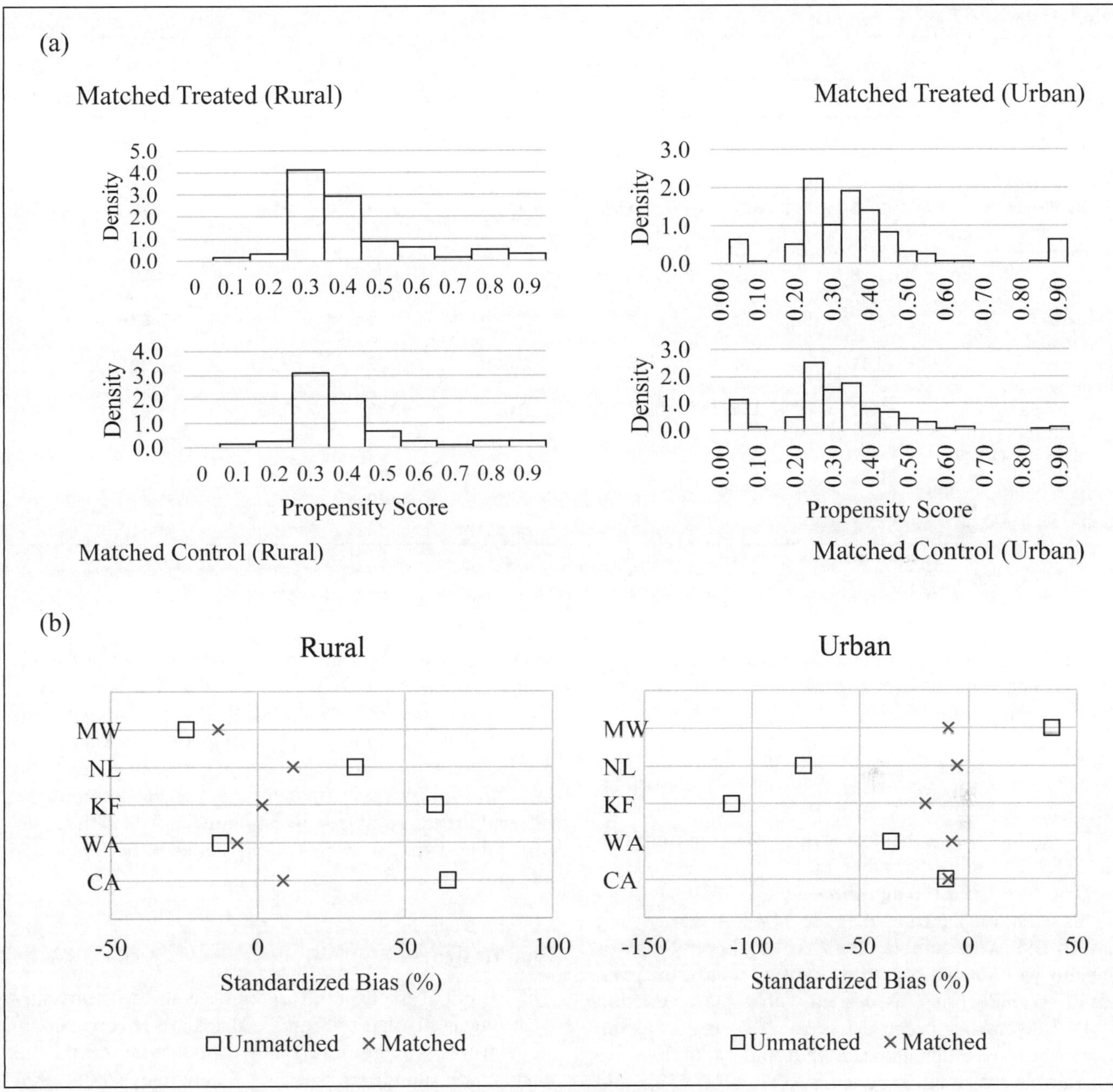

Figure 6. Showing (*a*) propensity score distribution, and (*b*) standardized bias of selected covariates.
Note: MW = median width; NL = number of lanes; KF = K-factor; CA=California; WA = Washington.

authors have inspected the standardized bias between the covariate means of two groups (Figure 6*b*). The standardized biases of the selected covariates seem to be very small; the highest value is less than 15%. This indicates that the treatment is independent of the covariates after the matching. Therefore, any observed differences between the treated and control groups will be due to the treatment variable.

Crash Analysis. The authors estimated negative binomial count data models using the matched data in order to assess the impact of rumble strips on the KA and B crash frequency. The model estimation and data matching occurred simultaneously and the authors then estimated a new model for each matched data. The final model is selected based on the quality of the matched data, goodness of fit of the models, and the residual tests. There are a total of four models per area type (eight models in total). The estimation results are reported in Table 7.

The coefficient estimates in this table indicate the expected impact of RS presence on crash types thus they

Table 7. Estimation Results

Parameters	MV-KA		SV-KA		MV-B		SV-B	
	Estimate	p value	Estimate	p value	Estimate	p value	Estimate	p value
Rural models								
Intercept	−1.03	<0.001	−1.60	<0.001	−0.33	<0.001	−0.55	<0.001
Rumble strips								
Coefficient	−0.56	<0.07	−0.58	0.11	−0.79	<0.001	−0.78	<0.001
% Decrease	43%		44%		55%		54%	
Confidence interval								
LL	−1.18		−1.26		−1.20		−1.20	
UL	−0.07		0.09		−0.38		−0.35	
Dispersion	0.44		0.27		0.51		0.69	
Urban models								
Intercept	−2.33	<0.001	−1.96	<0.001	−0.88	<0.001	−0.38	<0.001
Rumble strips								
Coefficient	−0.15	0.62	−0.20	0.62	−0.30	0.15	−0.25	0.29
% Decrease	14%		18%		26%		22%	
Confidence interval								
LL	−0.75		−0.98		−0.72		−0.73	
UL	0.44		0.58		0.10		0.21	
Dispersion	0.25		0.16		0.35		0.31	

Note: MV-KA = multiple vehicle fatal and injury crashes; SV-KA = single vehicle fatal and injury crashes; MV-B = multiple vehicle non-incapacitating injury crashes; SV-B = Single vehicle non-incapacitating injury crashes; LL = lower limit; UL = upper limit.

can be interpreted as the crash modification factors (CMF). The functional form of the CMF for the multiple-vehicle fatal and injury crashes (MV-KA) is

$$CMF_{RS,\,MV-KA} = (e)^{-0.56} \quad (6)$$

where

$CMF_{RS,\,MV-KA}$ = the rumble strip CMF for multiple-vehicle fatal and injury crashes, and

-0.56 = is the estimated coefficient.

This equation indicates that the MV-KA crash count on segments with rumble strips is 0.56 lower than for segments without rumble strips. In other words, the presence of rumble strips is associated with a 43% reduction in MV-KA crashes because $1-(e)^{-0.56}=0.43$. Overall the estimation results indicate that the rural freeways with rumble strips can experience 43%, 44%, 55%, and 54% less multiple-vehicle KA crashes, single-vehicle KA crashes, multiple-vehicle B (non-incapacitating) crashes, and single-vehicle B crashes compared with the segments without rumble strips. All of these results are significant. On the urban freeway segments, the study locations with rumble strips can experience 14%, 18%, 26%, and 22% fewer multiple-vehicle KA crashes, single-vehicle KA crashes, multiple-vehicle B crashes, and single-vehicle B crashes compared with the segments without the rumble strips although these results are not significant. The authors hypothesized that the insignificant effects could be due to the limited sample size of the treated segments.

The 95% confidence interval shows that the estimator will always be negative for severe crashes on rural freeways, i.e. on the rural segments the impact of rumble strips are expected to be mostly negative. The urban models did not show significant association with the rumble strips although according to the estimation results the rumble strips have a negative impact on the fatal and injury crash frequencies. The 95% confidence interval for the rumble strip estimator indicates the effect of rumble strips on severe crashes could be both negative and positive.

Discussions

The results of the exploratory analysis in this study support the claims that the rumble strips are effective countermeasures for preventing horizontal curve crashes as curves with rumble strips are less likely to experience KA crashes. Overall, both exploratory and causal data analysis indicate that rumble strips are associated with fewer fatal and injury crashes on both rural and urban freeways. The estimation results for the rural freeways were found to be significant and the confidence interval showed that the estimate for the rumble strip variable is stable since the signs did not change. The results for the urban freeways also indicated that installing rumble strips can result in fewer KA and B crashes; however, unlike the rural models, the rumble strip estimator for the urban models was not found to be significant. The authors assume that the insignificant results can be due to either the sample size of the urban segments with rumble strips or due to the fact that in urban areas the rumble strips do not play a significant role in fatal and injury crash occurrences.

Note that since the data used in this study is a cross-sectional data, the decreasing crash frequencies cannot be entirely attributed to the rumble strips. Nevertheless, the estimations are conducted using the matched data. Therefore, the decreasing crash counts should theoretically be attributed to the rumble strip presence.

Conclusions and Future Research

The authors developed this paper with the purpose of specifically evaluating the impact of rumble strips on freeway fatal and incapacitating injury crashes. The authors used data developed as part of the recent NCHRP 17-45 project. As indicated earlier the data used in this study is a cross-sectional data which is known for exhibiting selection bias. Therefore, in order to account for the selection bias, the authors have implemented the PSM method to evaluate the impact of rumble strips on fatal and injury crashes.

This study is expected to lead to additional research in order to draw definitive results. It is likely that future research on this topic could impact the HSM freeway and ramp CMF values.

Acknowledgments

The authors would like to thank the research team for NCHRP 17-45 and particularly Dr. James Bonneson for providing the dataset used in the study and for being available for the authors' questions related to this database.

Author Contributions

The authors confirm contribution to the paper as follows: study conception and design: Karen Dixon; data description: Bahar Dadashova and Lingtao Wu; analysis and interpretation of results: Bahar Dadashova (IRR, PSM, and Bootstrap Sampling), and Lingtao Wu (PSM); draft manuscript preparation: Bahar Dadashova. All authors reviewed the results and approved the final version of the manuscript.

References

1. Torbic, D. J., J. M. Hutton, C. D. Bokenkroger, K. M. Bauer, D. W. Harwood, D. K. Gilmore, J. M. Dunn, J. J. Ronchetto, E. T. Donnell, H. J. Sommer, P. M. Garvey, B. Persaud, and C. Lyon. *Guidance for the Design and Application of Shoulder and Centerline Rumble Strips*. Report NCHRP 641. Transportation Research Board, Washington, D.C., 2009.
2. Patel, R. B., F. M. Council, and M. S. Griffith. Estimating Safety Benefits of Shoulder Rumble Strips on Two-Lane Rural Highways in Minnesota - Empirical Bayes Observational before-and-after Study. *Transportation Research Record: Journal of the Transportation Research Board*, 2007. 2019: 205–211.
3. Khan, M., A. Abdel-Rahim, and C. J. Williams. Potential Crash Reduction Benefits of Shoulder Rumble Strips in Two-Lane Rural Highways. *Accident Analysis and Prevention*, Vol. 75, 2015, pp. 35–42.
4. Perrillo, K. *The Effectiveness and Use of Continuous Shoulder Rumble Strips*. Federal Highway Administration, Albany, New York, 1998.
5. Smith, E. B., and J. N. Ivan. Evaluation of Safety Benefits and Potential Crash Migration Due to Shoulder Rumble Strip Installation on Connecticut Freeways. *Statistical Methods; Highway Safety Data, Analysis, and Evaluation; Occupant Protection; Systematic Reviews and Meta-Analysis*, No. 1908, 2005, pp. 104–113.
6. Carrasco, O., J. McFadden, and P. Chandhok. Evaluation of the Effectiveness of Shoulder Rumble Strips on Rural Multi-Lane Divided Highways in Minnesota. Presented at 83rd Transportation Research Board Annual Meeting, Washington, D.C., 2004.
7. Torbic, D. J., J. M. Hutton, C. D. Bokenkroger, K. M. Bauer, E. T. Donnell, C. Lyon, and B. Persaud. Guidance on Design and Application of Rumble Strips. *Transportation Research Record: Journal of the Transportation Research Board*, 2010. 2149: 59–69.
8. Bonneson, J., S. Geedipally, M. Pratt, and D. Lord. *Safety Prediction Methodology and Analysis Tool for Freeways and Interchanges*. Report NCHRP 17-45. Texas Transportation Institute, College Station, Texas, 2012.
9. Hu, W., and E. T. Donnell. Median Barrier Crash Severity: Some New Insights. *Accident Analysis & Prevention*, Vol. 42, No. 6, 2010, pp. 1697–1704.
10. Hauer, E. Cause, Effect and Regression in Road Safety: A Case Study. *Accident Analysis & Prevention*, Vol. 42, No. 4, 2010, pp. 1128–1135.
11. American Association of State Highway and Transportation Officials. *Highway Safety Manual*. American Association of State Highway and Transportation Officials, Washington, D.C., 2010.
12. Rosenbaum, P. R., and D. B. Rubin. Reducing Bias in Observational Studies Using Subclassification on the Propensity Score. *Journal of the American Statistical Association*, Vol. 79, No. 387, 1984, pp. 516–524.
13. Sasidharan, L., and E. T. Donnell. Propensity Scores-Potential Outcomes Framework to Incorporate Severity Probabilities in the Highway Safety Manual Crash Prediction Algorithm. *Accident Analysis and Prevention*, Vol. 71, 2014, pp. 183–193.
14. Sasidharan, L., and E. T. Donnell. Application of Propensity Scores and Potential Outcomes to Estimate Effectiveness of Traffic Safety Countermeasures: Exploratory Analysis Using Intersection Lighting Data. *Accident Analysis and Prevention*, Vol. 50, 2013, pp. 539–553.
15. Guo, S., and M. W. Fraser. *Propensity Score Analysis*. Sage, Thousand Oaks, CA, 2014.

The Standing Committee on Highway Safety Performance (ANB25) peer-reviewed this paper (18-01440).

Article

TRR
JOURNAL OF THE TRANSPORTATION RESEARCH BOARD

Transportation Research Record
2018, Vol. 2672(30) 142–152
© National Academy of Sciences:
Transportation Research Board 2018
Article reuse guidelines:
sagepub.com/journals-permissions
DOI: 10.1177/0361198118788192
journals.sagepub.com/home/trr
SAGE

Crash Modification Factors for the Flashing Yellow Arrow Treatment at Signalized Intersections

Raghavan Srinivasan[1], **Bo Lan**[1], **Daniel Carter**[1], **Sarah Smith**[1], **and Kari Signor**[1]

Abstract

This paper presents the results of an evaluation of the flashing yellow arrow (FYA) treatment using data from signalized intersections in Nevada, North Carolina, Oklahoma, and Oregon. The evaluation method was an empirical Bayes before–after analysis. The treatments were divided into seven categories depending on the phasing system in the before period (permissive, protected–permissive, or protected), phasing system in the after period (FYA permissive or FYA protected–permissive), the number of roads where the FYA was implemented (one road or both roads), and the number of legs at the intersections (three or four). The first five treatment categories involved permissive or protected–permissive phasing in the before period. Intersections in these five treatment categories experienced a reduction in the primary target crashes under consideration: left turn crashes and left turn with opposing through crashes. The reduction ranged from 15% to 50%, depending on the treatment category. Intersections that had at least one protected left turn phase in the before period and had FYA protected–permissive left turn phase in the after period experienced an increase in left turn crashes and left turn with opposing through crashes, indicating that replacing a fully protected left turn with FYA will likely cause an increase in left turn crashes.

The primary intent of the flashing yellow arrow (FYA) for permissive left turn movements at signalized intersections is to help avoid confusion for drivers turning left on a permissive circular green signal. The concern is that drivers turning left on a permissive circular green signal indication might mistake that indication as implying the left turn has the right of way over opposing traffic, especially under some geometric conditions. This research examined the safety impacts of FYA using data from sites in Oklahoma, Oregon, Nevada, and North Carolina. The objective was to estimate the safety effectiveness of this strategy as measured by crash frequency. The primary target crash types are left turn crashes and left turn crashes with opposing through vehicles. However, changes in signal timing sometimes accompany changes in signal phasing, altering the green time that is available for through movements, and as a result, could affect the propensity for rear-end and angle crashes. For this reason, the evaluation included the following crash types:

- Total intersection crashes
- Intersection injury and fatal crashes (KABC)
- Intersection rear-end crashes (RE)
- Intersection angle crashes (ANG)
- Intersection left turn crashes (LT)
- Intersection left turn crashes with opposing through vehicles (LTOT)

The evaluation of overall effectiveness included the consideration of the installation costs and crash savings in terms of the benefit cost (BC) ratio.

Literature Review

Through the years, researchers have conducted different types of studies to evaluate FYA, including driver comprehension studies, driver simulator studies, operational studies, and crash-based studies. This literature review focuses on results from crash-based studies.

[1]University of North Carolina Highway Safety Research Center, Chapel Hill, NC

Corresponding Author:
Address correspondence to Raghavan Srinivasan: srini@hsrc.unc.edu

Perez conducted a simple/naïve before–after evaluation to compare the safety impacts of protected–permissive left turn (PPLT) FYA phase installation at seven intersections in the city of Federal Way in the state of Washington (*1*). The phasing in the before period included protected left turn phasing in some cases and traditional PPLT in other cases. Intersections that were converted from protected to PPLT FYA experienced an approximately 15% increase in crashes with a 41% increase in severity rates. On the other hand, intersections that were converted from traditional to PPLT to PPLT FYA experienced a 39% reduction in crashes and a 62% reduction in severity rates.

Srinivasan et al. investigated the safety impact of FYA installation using data from 51 signalized intersections in three states: North Carolina, Oregon, and Washington (*2*). As in the case of Perez, some intersection approaches had permissive left turn or traditional PPLT in the before period, while a majority of the converted approaches had protected left turn phase in the before period. Intersections that had permissive phasing in the before period experienced the largest reduction in crashes: a 36% reduction in intersection left turn crashes and a 25% reduction in total intersection crashes. Intersections with protected–permissive phasing in the before period experienced smaller reductions that were not statistically significant at the 95% confidence level: 19% reduction in intersection left turn crashes and an 8% reduction in total intersection crashes. Intersections with protected left turn phasing in the before period experienced a significant increase in crashes: 124% increase in intersection left turn crashes and a 34% increase in total intersection crashes.

Pulugurtha and Khader evaluated the impact of FYA installation at 18 signalized intersections in Charlotte (*3*). Not all legs underwent conversion to FYA. Similar to the Srinivasan et al. study, the before condition varied from permissive, traditional PPLT, and protected phasing. The researchers used data from the control intersections to estimate safety performance functions (SPFs) for an empirical Bayes (EB) before–after evaluation. Unlike some previous studies, left turning volume was available and included in the evaluation. Results indicated that left turning crashes involving left turning vehicles decreased in 14 out of the 18 intersections following FYA implementation. Benefits due to FYA installation were lower in higher volume sites with higher crash frequencies before implementation.

Simpson and Troy evaluated the impact of FYA using data from 222 signalized intersections in North Carolina (*4*). This before–after study used SPFs from Chapter 12 of the Highway Safety Manual (urban and suburban arterials) (*5*) to account for the changes in traffic volume from the before to the after periods. The authors estimated crash modification factors (CMFs) for total crashes, injury and fatal crashes, target crashes (collisions involving left turns with opposing through vehicles), and injury and fatal target crashes, on a treated approach. The authors identified target crashes based on a manual review of crash reports. As in the case of previous studies, the before phasing included permissive, traditional PPLT, and protected. The authors examined five categories of changes. Some intersections were exclusive to a particular category. However, in some intersections, the changes in one or more legs belonged to one category, while changes in one or more other legs belonged to another category. Two sets of results were reported for each category—the "first set" only included intersections that were exclusive to that category, and the "second set" included all legs that belonged to that category. The categories along with the CMFs are provided below:

- Category I. Permissive only to FYA PPLT; CMF of 0.598 for target crashes and 0.592 for target injury and fatal crashes.
- Category II. Protected only to FYA PPLT; CMF of 3.684 for target crashes and 4.778 for target injury and fatal crashes.
- Category IIA. Protected only to FYA PPLT with time-of-day (TOD) operation; CMF of 2.372 for target crashes and 2.371 for target injury and fatal crashes.
- Category III. Five-section doghouse PPLT to FYA PPLT; CMF of 0.838 for target crashes and 0.747 for target injury crashes.
- Category IV. Permissive only to FYA permissive only; CMF of 0.498 for target crashes and 0.349 for target injury crashes.

Consistent with the results from Srinivasan et al., Categories II and IIA that had protected left turn phasing in the before period experienced increases in target crashes. The sample in Category IIA was limited because only those time periods during FYA operation were included. Again, consistent with the results from Srinivasan et al., Category I, which included a phasing change to protected–permissive from permissive in addition to FYA implementation, was associated with larger reductions in target crashes. This is in comparison with Category III, which included a change to FYA without changes in the phasing. On the other hand, Category IV experienced a larger reduction in target crashes compared with the other categories, although this included a change to FYA without changes in the phasing.

Schattler et al., used the EB before–after approach to evaluate the safety impacts of FYA installation in the Peoria area of Illinois (*6*). The evaluation included 86 intersections where the state had implemented FYA on

164 approaches. The before condition at these intersections was the traditional PPLT with a five-section head, and the after condition was a FYA PPLT with a four-section head. A before–after EB evaluation was conducted. The evaluation focused on left turn–related and LTOT crashes. The study reported significant reductions for left turn and LTOT crashes:

- LT-related crashes at FYA approach CMF = 0.617, with a 95% confidence interval = 0.617 ± 0.012 = 0.605–0.629.
- LT-related crashes at FYA approach with supplemental sign CMF = 0.589, with a 95% confidence interval = 0.589 ± 0.016 = 0.573–0.605.
- LTOT crashes at FYA approach CMF = 0.714, with a 95% confidence interval = 0.714 ± 0.016 = 0.698–0.730.
- LTOT crashes at FYA approach with supplemental sign CMF = 0.711, with 95% confidence interval = 0.711 ± 0.024 = 0.687–0.735.

The study reported a BC ratio of 19.8 to 1 for the FYA treatment.

Summary of Literature Review

Previous studies have generally shown that the FYA treatment is associated with a reduction in left turn crashes as long as the before period phasing is not a protected left turn. The specific CMF values do vary for many reasons including:

- The methodology used for the evaluation
- The definition of target crashes
- The phasing in the before period
- The states used in the evaluation

Most of the evaluations used data from only one state. Srinivasan et al. used data from three states, but the sample was limited to 51 sites with a majority of sites (29 sites) having protected left turn phasing in the before period (*2*). It is clear that an evaluation with a large sample of sites from multiple states would provide useful information to the practitioner on the effectiveness of this treatment under different circumstances.

Overview of Methodology

This evaluation used the EB methodology for observational before–after studies. This method is currently well established in the field of highway safety. It accounts for the possible bias due to regression to the mean (RTM) and properly accounts for changes in traffic volume during the study period (*7*).

The following steps are needed to conduct an EB before–after evaluation:

1. Identify a reference group of sites without the treatment, but similar to the treatment sites in terms of the major factors that affect crash risk including traffic volume and other site characteristics.
2. Estimate SPFs using data from the reference sites relating crashes to independent variables, such as traffic volume and other site characteristics. In some cases, for example, when the treatment is implemented systemwide or if the bias due to RTM is considered minimal, the before data from the treatment sites is used along with reference or comparison sites to estimate the SPFs (*8*). In fact, in this evaluation, the before data from the treatment sites were combined with the reference sites for estimating SPFs.
3. In estimating SPFs, calibrate annual SPF multipliers to account for the temporal effects (e.g., variation in weather, demography, and crash reporting) on safety. The annual SPF multiplier is the ratio of the observed crashes to the predicted crashes from the SPF. In using the annual SPF multipliers from the SPFs to account for temporal effects, it is assumed that the trends in the crash counts are similar in the treatment and reference groups.
4. Use the SPFs, annual SPF multipliers, and data on traffic volumes for each year in the before period for each treatment site to estimate the number of crashes that would be predicted for the before period in each site.
5. Calculate the EB estimate of the expected crashes in the before period at each treatment site as the weighted sum of the actual crashes in the before period and predicted crashes from Step 4.
6. For each treatment site, estimate the product of the EB estimate of the expected crashes in the before period and the SPF predictions for the after period divided by the SPF predictions for the before period. This is the EB expected number of crashes that would have occurred had there been no treatment. The variance of this expected number of crashes is also estimated in this step. The expected number of crashes without the treatment along with the variance of this parameter and the number of reported crashes after the treatment is used to calculate the CMF and the standard deviation of the CMF.

When planning a before–after safety evaluation study, it is vital to ensure that enough data are included to

Table 1. Treatment Categories

Treatment category	Before phasing	After phasing	Number of legs	Number of treatment sites				
				NV	NC	OK	OR	Total
1	Traditional PPLT	FYA PPLT on one road	3	3	25	8	4	40
2	Traditional PPLT	FYA PPLT on one road	4	48	55	14	19	136
3	Traditional PPLT	FYA PPLT on both roads	4	33	6	23	2	64
4	Permissive or traditional PPLT	FYA permissive on one road	4	0	25	0	0	25
5	Permissive	FYA permissive on one road	4	0	12	0	0	12
6	At least one protected phase	FYA PPLT without TOD	4	0	11	6	1	18
7	At least one protected phase	FYA PPLT with TOD	4	0	12	0	0	12

Note: PPLT = protected–permissive left turn operation; TOD = time-of-day operation, where the signal operates as FYA only during off-peak hours and protected phase during peak hours.

Table 2. Summary Statistics for Treatment Sites

Variable	Mean values by treatment category						
	1	2	3	4	5	6	7
Before years/site	6.20	5.54	5.97	7.20	7.58	6.20	7.33
After years/site	2.85	2.61	1.59	2.80	2.42	2.85	2.67
AADT on major road (before)	17,079	19,487	20,551	19,599	14,614	17,079	28,303
AADT on major road (after)	16,115	18,473	19,942	18,641	15,950	16,115	26,150
AADT on minor road (before)	5,035	6,322	8,831	5,344	4,841	5,035	18,381
AADT on minor road (after)	5,014	5,897	8,529	4,817	5,470	5,014	16,670
Total crashes (before)/site-year	3.70	6.31	9.07	5.86	5.49	3.70	18.10
Total crashes (after)/site-year	3.01	5.47	7.14	5.62	6.19	3.01	17.13
Injury crashes (before)/site-year	1.36	2.40	3.55	2.14	2.54	1.36	5.52
Injury crashes (after)/site-year	1.09	2.09	2.88	1.74	2.35	1.09	6.01
RE crashes (before)/site-year	1.44	2.39	3.19	2.22	1.67	1.44	9.40
RE crashes (after)/site-year	1.25	2.15	2.90	2.14	2.68	1.25	7.30
ANG crashes (before)/site-year	0.59	1.69	3.07	1.32	1.00	0.59	2.78
ANG crashes (after)/site-year	0.41	1.51	2.24	1.31	0.86	0.41	3.05
LT crashes (before)/site-year	0.76	1.60	3.15	1.07	1.77	0.76	1.05
LT crashes (after)/site-year	0.64	1.28	1.85	0.75	1.29	0.64	1.60
LTOT crashes (before)/site-year	0.57	1.06	2.35	0.75	1.31	0.57	0.83
LTOT crashes (after)/site-year	0.52	0.72	1.32	0.53	0.92	0.52	1.08

Note: AADT = annual average daily traffic.

statistically detect the expected change in safety. Even though at the planning stage, the expected change in safety is not known, it is still possible to make a rough determination of how many sites the study will require, based on the best available information about the expected change in safety. For a detailed explanation of sample size considerations, as well as estimation methods, see Chapter 9 of Hauer (*7*). In this study, the methods from Chapter 9 of Hauer (*7*) were used to determine the number of sites that are needed to statistically detect the expected change in safety.

Data and Treatment Categories

Nevada, North Carolina, Oklahoma, and Oregon provided data for this study. These states also provided data on details about the intersections, traffic volumes on the major and minor roads, and crashes for both installation and reference sites. The evaluation included 307 treated sites and 438 reference sites from these four states.

As mentioned earlier, previous research has revealed that the before condition could have a significant impact on safety. For example, if FYA is introduced as a replacement for protected left turn phasing, it could lead to an increase in left turn crashes. For this reason, based on the phasing system before and after the implementation of the FYA, seven treatment categories were identified. Table 1 shows the seven treatment categories along with the number of treated intersections from each state in each category.

Table 2 shows the summary statistics for the treatment sites. Table 3 shows the summary statistics for the reference sites. Further details about the data, including summary statistics by state will be available in an upcoming FHWA report (*9*).

Table 3. Summary Statistics for Reference Sites

	Mean values by number of intersection legs	
Variable	3-leg intersections (48 sites)	4-leg intersections (390 sites)
Number of years	8.42	7.48
AADT (major)	19,525	21,331
AADT (minor)	7,997	9,231
Total crashes/site-year	4.56	9.01
Injury crashes/site-year	1.81	3.96
RE crashes/site-year	1.84	3.62
ANG crashes/site-year	1.30	3.52
LT crashes/site-year	1.20	2.48
LTOT crashes/site-year	0.60	1.14

SPFs

The EB methodology uses SPFs to estimate the safety effectiveness of this strategy. Generalized linear modeling was used to estimate model coefficients assuming a negative binomial error distribution, which is consistent with the state of research in developing these models. The independent variables included the following:

- Major road AADT
- Minor road AADT
- Number of legs (three or four legs; this is a categorical variable)
- Maximum left turn protection (based on the maximum left turn protection at an intersection–protected, protected–permissive, or permissive)
- Number of through lanes on the major road
- Presence/absence of median on the major road
- Number of approaches with left turn lanes

The variables are included in a log-linear form as follows:

$$Y = \exp\,(a_0 + a_1X_1 + a_2X_2 + \cdots ..a_nX_n)$$

Where,
a_0 = the intercept
a_1 through a_n = the coefficients for independent variables X_1 through X_n
Y = number of predicted crashes from the SPF

The project team estimated separate SPFs for each crash type for each state. After estimating the SPFs, the project team estimated annual SPF multipliers to account for temporal effects. The SPFs are not included in this paper for the sake of brevity, but are available in an upcoming FHWA report (*9*).

Results

This section presents the CMFs that were estimated based on the EB procedure (discussed in the Overview of Methodology section) for Treatment Categories 1–7. Each table shows the crash type, the number of crashes in the after period, the expected number of crashes had the treatment not been implemented, the CMF, and the standard error of the CMF. KABC crash type represents the effect on crashes that are fatal (K) or any kind of injury (i.e., A, B, or C injury). CMFs statistically different from 1.0 at the 95% confidence level are indicated with an asterisk (*).

Results for Conversion of Traditional PPLT to FYA PPLT

By far the most common type of FYA implementation was converting a traditional PPLT signal to a FYA PPLT signal (see installation numbers in Table 1). The states and cities varied slightly in how this conversion was undertaken. The results in Table 4 provide the CMFs according to the manner of FYA conversion. Some agencies implemented the conversion only on one road of the intersection (Categories 1 and 2), whereas others implemented it for both roads (Category 3). The safety effects of FYA in these three categories are similar. Total crashes decreased between 10% and 20% and injury and fatal crashes decreased about 20%. Left turn crashes and left turn opposing through crashes (the target crash types for this countermeasure) decreased between 25% and 50%, an effect that was statistically significant for the four-leg intersections but not the three-leg intersections. Category 2 sites also saw a reduction of 12% in rear-end crashes. There was no significant reduction to angle crashes.

Results for Conversion of Permissive or Traditional PPLT to FYA Permissive

Categories 4 and 5 represent conversions from permissive *or* traditional PPLT where the after period is a FYA permissive phase. The results for these two categories (shown in Table 5) are fairly consistent with each other, though larger reductions were seen when the signal phasing before FYA was only permissive. Injury and fatal crashes decreased by approximately 20% in both categories. Left turn crashes were reduced by 27–39%, and left turn with opposing through crashes were reduced by 27–45%.

Results for Conversion of Protected to FYA PPLT

Some agencies converted at least one intersection approach from fully protected left turn phasing to FYA PPLT (Categories 6 and 7). Agencies typically make this

Table 4. CMF Results for Traditional PPLT to FYA PPLT

Category	Crash type	Actual after	Expected after	CMF	SE of CMF
Category 1 (traditional PPLT to FYA PPLT on one road, three legs)	Total	363	427.2	0.849*	0.053
	KABC	129	162.7	0.791*	0.080
	RE	148	169.4	0.871	0.084
	ANG	49	63.5	0.768	0.122
	LT	80	99.0	0.804	0.106
	LTOT	60	70.4	0.846	0.131
Category 2 (traditional PPLT to FYA PPLT on one road, three legs)	Total	1951	2,194.8	0.889*	0.027
	KABC	722	900.3	0.801*	0.038
	RE	753	851.4	0.884*	0.042
	ANG	486	505.4	0.960	0.054
	LT	413	552.9	0.746*	0.047
	LTOT	200	324.1	0.615*	0.055
Category 3 (traditional PPLT to FYA PPLT on both roads, four legs)	Total	750	916.4	0.818*	0.036
	KABC	286	365.3	0.782*	0.055
	RE	306	338.6	0.902	0.066
	ANG	207	233.7	0.885	0.068
	LT	185	296.2	0.624*	0.053
	LTOT	75	147.6	0.507*	0.064

*Statistically different from 1.0 at the 95% confidence level.

Table 5. CMF Results for Permissive or Traditional PPLT to FYA Permissive

Category	Crash type	Actual after	Expected after	CMF	SE of CMF
Category 4 (permissive or traditional PPLT to FYA permissive on one road)	Total	409	410.0	0.997	0.058
	KABC	124	153.1	0.808*	0.082
	RE	159	157.9	1.005	0.093
	ANG	94	90.9	1.030	0.123
	LT	55	75.1	0.729*	0.109
	LTOT	39	52.9	0.733*	0.130
Category 5 (permissive to FYA permissive on one road)	Total	192	209.3	0.915	0.078
	KABC	71	93.6	0.787*	0.104
	RE	84	68.0	1.227	0.165
	ANG	23	30.2	0.753	0.173
	LT	42	68.2	0.612*	0.105
	LTOT	30	54.3	0.548*	0.111

*Statistically different from 1.0 at the 95% confidence level.

change to improve capacity by allowing more time for through movements, but the implications on safety are important to recognize. Not surprisingly, both types of left turn crashes increased in these situations (see Table 6). This result was true whether the agency implemented FYA with or without TOD operation, though the effect was statistically significant for only those sites without TOD operation. The significant results showed that all left turn crashes increased by 55% and left turn with opposing through crashes increased 91%.

Results for Conversion of Traditional PPLT to FYA PPLT by State

Since converting a traditional PPLT signal to FYA PPLT at four-leg intersections (Categories 2 and 3) was the most common implementation of FYA, Tables 7 and 8 provide further results from these categories by state. In Category 2, sites in Oregon and North Carolina experienced a statistically significant reduction in left turn crashes following the treatment. In Category 3, the sites in all the states experienced a statistically significant reduction in left turn crashes following the treatment. However, the magnitude of the reduction was not consistent across the states. The results from Oklahoma are a bit surprising especially with regard to the increase in rear-end and angle crashes following the implementation of the FYA. Part of this may be due to the differences in FYA implementation in Oklahoma compared with the other states. The FYA installation sites in Oklahoma were located in a single city (Norman, Oklahoma), where the city had conducted a citywide installation by

Table 6. CMF Results for Protected to FYA PPLT

Category	Crash type	Actual after	Expected after	CMF	SE of CMF
Category 6 (at least one protected phase to FYA PPLT without TOD)	Total	378	359.1	1.051	0.065
	KABC	120	118.3	1.011	0.110
	RE	152	164.0	0.925	0.087
	ANG	57	55.8	1.014	0.159
	LT	82	52.5	1.551*	0.219
	LTOT	71	36.8	1.910*	0.299
Category 7 (at least one protected phase to FYA PPLT with TOD)	Total	518	531.6	0.974	0.050
	KABC	178	163.1	1.089	0.095
	RE	227	250.9	0.903	0.068
	ANG	96	81.8	1.169	0.141
	LT	44	34.4	1.267	0.226
	LTOT	30	25.7	1.151	0.242

*Statistically different from 1.0 at the 95% confidence level.

Table 7. CMFs by State for Treatment Category 2

State (number of sites)	Crash type	Actual after	Expected after	CMF	SE of CMF
Oklahoma (14)	Total	88	77.6	1.131	0.135
	KABC	26	28.0	0.921	0.195
	RE	55	33.3	1.641*	0.255
	ANG	11	8.6	1.253	0.402
	LT	17	21.0	0.804	0.207
	LTOT	14	19.1	0.725	0.204
Oregon (19)	Total	246	339.7	0.719*	0.075
	KABC	125	173.8	0.711*	0.099
	RE	97	167.5	0.572*	0.086
	ANG	32	44.9	0.694	0.164
	LT	71	120.8	0.577*	0.103
	LTOT	45	75.0	0.584*	0.127
Nevada (48)	Total	430	447.4	0.960	0.052
	KABC	188	199.4	0.941	0.077
	RE	159	181.8	0.873	0.077
	ANG	167	164.6	1.013	0.087
	LT	142	120.5	1.176	0.112
	LTOT	–	–	–	–
North Carolina (55)	Total	1,187	1,330.1	0.892*	0.033
	KABC	383	499.0	0.767*	0.047
	RE	442	468.8	0.942	0.056
	ANG	276	287.3	0.959	0.074
	LT	183	290.6	0.628*	0.055
	LTOT	141	229.9	0.611*	0.061

Note: – = crash data not provided by state.
*Statistically different from 1.0 at the 95% confidence level.

converting all PPLT signals to FYA, rather than targeting specific sites. However, the full reasons for the differences in the results from the individual states are not known at this time.

Crash Modification Functions

Crash modification functions (CMFunctions) were estimated to investigate the effect of site characteristics on the effectiveness of the FYA implementation. CMFunctions were estimated only for Categories 2 and 3, the two most common categories, which had sufficient sample size to provide useful results.

The traditional approach for estimating CMFunctions includes the use of the CMF value as the dependent variable and site/treatment characteristics as independent variables:

$$CMF = f(site\ characteristics)$$

Table 8. CMFs by State for Treatment Category 3

State (number of sites)	Crash type	Actual after	Expected after	CMF	SE of CMF
Oklahoma (23)	Total	299	286.9	1.041	0.069
	KABC	102	105.9	0.961	0.107
	RE	127	111.1	1.140	0.115
	ANG	50	25.8	1.921*	0.319
	LT	68	103.2	0.657*	0.087
	LTOT	60	99.4	0.602*	0.084
Oregon (2)	Total	38	64.7	0.552*	0.156
	KABC	18	30.5	0.525*	0.198
	RE	15	37.8	0.357*	0.135
	ANG	9	8.5	0.897	0.412
	LT	6	20.0	0.247*	0.126
	LTOT	2	5.6	0.226*	0.149
Nevada (33)	Total	268	353.3	0.758*	0.050
	KABC	138	162.8	0.846	0.078
	RE	90	115.3	0.779*	0.088
	ANG	127	162.7	0.780*	0.074
	LT	95	117.3	0.809*	0.089
	LTOT	–	–	–	–
North Carolina (6)	Total	145	211.5	0.684*	0.065
	KABC	28	66.0	0.421*	0.086
	RE	74	74.4	0.989	0.137
	ANG	21	36.8	0.565*	0.136
	LT	16	55.7	0.285*	0.074
	LTOT	13	42.6	0.303*	0.088

Note: – = crash data not provided by state.
*Statistically different from 1.0 at the 95% confidence level.

Where, f represents a mathematical function. This CMFunction could then be estimated as a regression equation. In order to estimate the CMF using a regression equation, reliable estimates of CMFs and their variances (which typically serve as a weight) are necessary. In order to have reliable estimates of these parameters, researchers often combine sites with similar characteristics, especially if there are sites where the reported crashes in the after period are 0. However, this aggregation can lead to loss of useful information. In this study, a different approach was used to overcome the disadvantage of losing information due to aggregation. This approach involves rewriting the CMFunction as follows:

$$CMF_i = \frac{\delta_i}{\pi_i} = f(site\ characteristics)$$

Where,
$\pi_i =$ expected number of crashes that would have occurred in the after period without the strategy at site i
$\delta_i =$ number of reported crashes in the after period at a particular site i

Using fundamental algebra, the CMFunction above can be rewritten as follows:

$$\delta_i = \pi_i \times f(site\ characteristics)$$

Written in this form, it is possible to estimate this equation as a count data model (such as Poisson or negative binomial model) with δ as the dependent variable and π as the offset. This new approach for estimating CMFunctions has been found to have some advantages compared with the traditional approach (*10*). One limitation with this approach is that the offset is an estimated value from the EB procedure with a variance, and therefore, not a fixed measure value. There has been some limited research in traffic safety on the implications of such errors/variance in the independent variables, but further research is necessary (e.g., *11*).

CMFunctions were estimated for the two target crash types: left turn crashes and left turn with opposite direction crashes. The CMFunctions for Category 3 did not provide useful results. Similarly, the CMFunctions for left turn crashes for Category 2 were not useful. The CMFunction for LTOT crashes for Category 2 was the following:

$$CMF = 0.694 \times (Exp\ bef\ per\ year)^{-0.2626}$$

Where,

Exp bef per year = the EB expected LTOT crashes per year at the intersection level in the before period (i.e., before the FYA was implemented)

Table 9. Results of Economic Analysis

Treatment category	Economic benefits due to crash reduction (per intersection per year) (mean)[a]	Annualized treatment cost (per intersection per year)	BC ratio (mean)	BC ratio (min, max)[b]
Category 1	$ 72,010	$854	84:1	(46:1, 116:1)
Category 2	$ 117,626	$1,709	69:1	(38:1, 95:1)
Category 3	$ 191,990	$3,417	56:1	(31:1, 78:1)
Category 4	$ 245,410	$1,709	144:1	(79:1, 198:1)
Category 5	$ 152,535	$1,709	89:1	(49:1, 123:1)

[a]The low and high values for the economic benefits were estimated by multiplying the mean values by 0.55 and 1.38, respectively.
[b]The min value of the BC ratio corresponds to the low value of the economic benefit, and the max value of the BC ratio corresponds to the high value of the economic benefit.

Since the power for *Exp bef per year* is negative, the treatment is expected to be more effective in locations that have higher expected LTOT crashes before the treatment is implemented. This function means that a multiplicative change of X in *Exp bef per year* would result in a multiplicative change of $X^{-0.2626}$ in the CMF value. For example, for every twofold multiplicative increase in *Exp bef per year*, the CMF experiences a multiplicative reduction by a factor of $2^{-0.2626}$ (=0.83), that is, the LTOT crashes are expected to reduce by 17%. In the data that were used to estimate the CMFunction, the *Exp bef per year* varied from 0.14 to 5.07. For *Exp bef per year* below about 0.25, CMFs are unexpectedly higher than 1.0. However, there are only 14 sites with *Exp bef per year* between 0.14 and 0.25, and the results could be skewed due to a few outliers. For practical purposes, it is reasonable to apply this CMFunction only for *Exp bef per year* higher than 0.25.

Economic Analysis

Economic analyses were undertaken for Treatment Categories 1–5. Treatment Categories 6 and 7 (change from protected to FYA PPLT) are implemented for reasons other than safety, and as a result, were not included in the economic analysis.

For the benefit calculations, the most recent FHWA mean comprehensive crash costs by crash severity were used (*12*). These costs, based on 2001 crash costs, were updated to 2015 dollars by applying the ratio of the U.S. DOT 2015 value of a statistical life of $9.4 million to the 2001 value of $3.8 million (*13*). Applying this ratio resulted in an aggregate 2015 unit cost of $227,744 for injury and fatal crashes and $17,513 for PDO crashes. Sensitivity analyses were conducted, which involved multiplying the mean value of the economic benefits by 0.55 to get a low estimate of the benefits, and multiplying the mean value of the economic benefits by 1.38 to get a high estimate of the benefits.

For treatment cost, Oklahoma indicated that the installation cost was about $6,500 for a four-leg intersection, that is, about $1,625 per approach leg. In a recent study in Illinois (*6*), installation cost was assumed to be $6,000 per approach leg. The higher installation cost from Illinois (i.e., $ 6,000 per approach leg) was assumed to get a conservative estimate for the BC ratio. In using this cost, it was assumed that the signal pole is structurally adequate to accommodate the retrofit. In addition, it was also assumed that there would be no additional maintenance costs for the FYA compared with traditional phasing systems. For these calculations, the discount rate (as a decimal) was assumed to be 0.07, and the expected service life (years) was assumed to be 10 years. The BC ratio was calculated as the ratio of the annual crash savings to the annualized treatment cost. Table 9 provides the results.

Conclusion

Data from four states were used to evaluate the safety of the FYA treatment. The treatments were divided into seven categories depending on the phasing system in the before period (permissive, protected–permissive, or protected), phasing system in the after period (FYA permissive or FYA protected–permissive), the number of roads where the FYA was implemented (one road or both roads), and the number of legs at the intersections (three or four). The first five categories involved permissive or protected–permissive phasing in the before period. Intersections in these five treatment categories experienced a reduction in the primary target crashes under consideration: left turn crashes and left turn with opposing through crashes. The reduction ranged from 15% to 50%, depending on the treatment category. Intersections in Categories 6 and 7 had at least one protected left turn phase in the before period, and after phasing had FYA protected–permissive left turn phase without TOD operation (Category 6) and with TOD operation (Category 7). Consistent with results from previous studies, these intersections with fully protected phasing in the before period experienced an increase in left turn and left turn with opposing through crashes. The study also produced a function that relates the CMF value to the EB expected LTOT crashes per year at the intersection level in the before period.

There are some limitations in this study. Left turn volumes were not available and, as a result, the evaluation could not include them. The evaluation focused on intersection level crashes instead of approach level crashes, because it was not possible to reliably obtain approach level crashes from the coded electronic crash reports.

Recommendations for Implementation of Results

This section provides the CMFs recommended for use by practitioners who are considering implementing FYA. Only CMFs that are statistically different from 1.0 at the 95% confidence level for total crashes, KABC crashes, left turn crashes, and left turn opposing through movement crashes, are mentioned here.

Conversion of traditional PPLT to FYA PPLT on one road at a three-leg intersection:

- Total crashes: CMF = 0.849
- KABC crashes: CMF = 0.791
- Left turn crashes and left turn opposing through movement crashes: CMFs were not statistically significant

Conversion of traditional PPLT to FYA PPLT on one road at a four-leg intersection:

- Total crashes: CMF = 0.889
- KABC crashes: CMF = 0.801
- All left turn crashes: CMF = 0.746
- Left turn opposing through movement crashes: CMF = 0.615

Conversion of traditional PPLT to FYA PPLT on both roads at a four-leg intersection:

- Total crashes: CMF = 0.818
- KABC crashes: CMF = 0.782
- All left turn crashes: CMF = 0.624
- Left turn opposing through movement crashes: CMF = 0.507

Conversion of permissive or traditional PPLT to FYA permissive on one road at a four-leg intersection:

- Total crashes: CMF not statistically significant
- KABC crashes: CMF = 0.808
- All left turn crashes: CMF = 0.729
- Left turn opposing through movement crashes: CMF = 0.733

Conversion of permissive to FYA permissive on one road at a four-leg intersection:

- Total crashes: CMF not statistically significant
- KABC crashes: CMF = 0.787
- All left turn crashes: CMF = 0.612
- Left turn opposing through movement crashes: CMF = 0.548

Conversion of protected left turn to FYA PPLT (without TOD operation):

- Total crashes and KABC crashes: CMF were not statistically significant
- All left turn crashes: CMF = 1.551
- Left turn opposing through movement crashes: CMF = 1.910

Acknowledgments

This paper is based on a study funded by the Federal Highway Administration, Office of Safety Research and Development, under Contract DTFH61-13-D-0001. The FHWA Contracting Officer's Representative for this project was Roya Amjadi. The authors gratefully acknowledge the assistance of: Nevada DOT; North Carolina DOT; Oklahoma DOT; City of Norman, Oklahoma; City of Edmond, Oklahoma; and Oregon DOT.

Author Contributions

The authors confirm contributing to the paper as follows: study conception and design: Raghavan Srinivasan, Daniel Carter, and Bo Lan; data collection: Daniel Carter, Sarah Smith, and Kari Signor; analysis and interpretation of results: Bo Lan, Raghavan Srinivasan, and Daniel Carter; draft manuscript preparation: Raghavan Srinivasan, Daniel Carter, and Bo Lan. All authors reviewed the manuscript and approved the final version of the manuscript.

References

1. Perez, R. A. *Safety Implications of Conversions to Flashing Yellow Arrow Indications*. City of Federal Way, Washington, 2010.
2. Srinivasan, R., J. Baek, S. Smith, C. Sundstrom, D. Carter, C. Lyon, B. Persaud, F. Gross, K. Eccles, A. Hamidi, and N. Lefler. *NCHRP Report 705: Evaluation of Safety Strategies at Signalized Intersections, Appendix E: Evaluation of Flashing Yellow Arrow for Permissive Left Turn Movements at Signalized Intersections*. Transportation Research Board of the National Academies, Washington, D.C., 2011.
3. Pulugurtha, S. S., and K. S. C. Khader. Assessing the Effect of Introducing a Permitted Phase Through the Use

of Flashing Yellow Arrow Signal for Left-Turning Vehicles. *International Journal of Injury Control and Safety Promotion*, Vol. 21, No. 4, 2014, pp. 338–347.
4. Simpson, C. L., and S. A. Troy. Safety Effectiveness of Flashing Yellow Arrow: Evaluation of 222 Signalized Intersections in North Carolina. *Transportation Research Record: Journal of the Transportation Research Board*, 2015. 2492: 46–56.
5. AASHTO. *Highway Safety Manual*. AASHTO, 2010. http://www.highwaysafetymanual.org/Pages/default.aspx.
6. Schattler, K. L., E. Anderson, and T. Hanson. *Safety Evaluation of Flashing Yellow Arrows for Protected/Permissive Left-Turn Control*. Research Report FHWA-ICT-16-010. Illinois Center of Transportation Center, 2016.
7. Hauer, E. *Observational Before-After Studies in Road Safety*. Pergamon Press, Oxford, UK, 1997.
8. Persaud, B., and C. Lyon. Empirical Bayes Before-After Safety Studies: Lessons Learned from Two Decades of Experience and Future Directions. *Accident Analysis and Prevention*, Vol. 39, No. 3, 2007, pp. 546–555.
9. Srinivasan, R., B. Lan, D. Carter, S. Smith, and K. Signor. *Safety Evaluation of the Flashing Yellow Arrow Treatment at Signalized Intersections*. Federal Highway Administration, Washington, D.C., 2018.
10. Srinivasan, R., and B. Lan. *Estimation of Crash Modification Functions Using Site-Level Information from Results of Empirical Bayes Before-After Evaluations*. Submitted to Southeastern Transportation Center, 2016.
11. Wright, D. E., and V. Barnett. *Fitting Predictive Accident Models in GLIM with Uncertainty in Flow Estimates*. Contractor Report 286. Transport and Road Research Laboratory, 1991.
12. Council, F., E. Zaloshnja, T. Miller, and B. Persaud. *Crash Cost Estimates by Maximum Police-Reported Injury Severity within Selected Crash Geometries*. FHWA-HRT05-051. Federal Highway Administration, McLean, Va., 2005.
13. U.S. Department of Transportation. *Guidance on Treatment of the Value of a Statistical Life (VSL) in U.S. Department of Transportation Analyses—2015 Adjustment*. U.S. Department of Transportation, Memorandum, 2015. https://cms.dot.gov/sites/dot.gov/files/docs/VSL2015_0.pdf. Accessed May 10, 2016.

The Standing Committee on Highway Safety Performance (ANB25) peer-reviewed this paper (18-01776).

Sage Publications USA
sagepub.com

ISBN 978-1-5108-7833-4

9 781510 878334

CURRAN ASSOCIATES INC.
proceedings
.com